Exam Practice
A LEVEL

A Level
Exam Practice

Covers AS and A2

Geography

Authors

Peter Goddard
John Snelling

Contents

AS and A2 exams

Different types of questions

At AS and also at A2 the three main types of assessment are:

Structured questions

- These questions are sub-divided.
- Expect the questions to increase in Difficulty as it progresses.
- Based on a support resource such as a diagram, photograph or map.
- There is a focus on description rather than explanation. You need to be concise.
- You must try to fit answers to the lines offered and to the mark weighting. Every word you write counts!
- Practising such questions is an absolute must.
- You will probably have an opportunity to use exemplars and to include ideas from your fieldwork.

Data response questions

- These questions attempt to test especially your geographical skills.
- These questions are different to structured questions in that the standard expected is probably a little higher. Precision in description is expected, and some elementary analysis might be called for.
- You might have to summarise a table, label (annotate) a diagram or photo or describe in a more technical sort of way

Extended prose

- This type of question requires you to write several sentences or up to a page in your answer.
- Such questions attempt to assess your ability to communicate ideas clearly and logically.
- In these questions your answer must be packed with facts and you must know your examples. But above all you must respond to command words.
- You can use diagrams sparingly. Ensure they contribute to the answer!
- Some planning, as in all written assessments, pays dividends here.

At A2, because candidates are expected to have a deeper understanding, greater knowledge and to be able to make links between different parts of their specification it is expected these candidates can handle the more complex technique of essay writing. A2 essays are a step up from extended prose therefore one might rightly expect there to be an incline in the complexity and type of language and command words used. The lists below attempt to show how the step up is made between AS and A2; there are of course many other command words used.

AS:

- Describe: details of appearance and characteristics needed.
- How: process or mechanism recognition.
- Why: explain.
- State: briefly, perhaps one word?
- Illustrate: with examples and case studies

A2:

In addition to the words to the left it also uses:

- Examine: that is investigate in detail.
- Criticise: perhaps explores weaknesses in arguments.
- Discuss: considers both sides of an argument.
- Justify: an argument in support of a particular view.
- Comment: a balanced view or judgement.

Successful revision

What Examiners look for

- In A2 essay work especially, utilise every word that you put down, display knowledge and understanding, use case studies, respond to command words, be organised, balanced and evaluative and offer a conclusion. Most A2 essays will look for a 40–45 minute answer.

- Primarily examiners are looking for answers that are clear, legible and concise.

- The mark schemes examiners use enable candidates to offer a wide range of interpretations to questions. They are not looking for a specific or pre-determined answer.

- Diagram used must make a real contribution. They have to be both neat and accurate.

- Don't over-run/over-write, you get no extra marks for doing so. Check the quality of your written work.

What makes an A, C and E grade candidate?

You will want to get the best possible grade you can. The way to do this is to have a good all-round understanding and knowledge of Geography.

- **A Grade candidates**, will perform strongly in all modules. A wide base of knowledge and understanding will be obvious and well applied in the assessments. Minimum mark for an A grade is 80%.

- **C Grade candidates**, will have a reasonable knowledge of geography, but they may be hesitant or unsure about applying their knowledge. Modular performances will have been variable. Minimum marks for a C grade is 60%.

- **E Grade candidates**, have a poor knowledge of geography and find great difficulty applying their knowledge effectively in unusual situations. The minimum mark for an E grade candidate is 40%.

Revision

Revision is the most important part of your preparations for both the AS and A2 examinations.

Here's how we suggest you revise:

- Ensure you know the requirements of your specification; what content do you have to understand, what attributes and values are asked for and what skills are needed?

- Know what type of paper is set and how it is organised.

- Some weeks before the examinations begin new teaching from the specification should cease and revision in school or college will begin. Have your programme of revision running before the school, revision sessions begin, this allows problems, misinterpretations and difficulties to be sorted out easily as they crop up. And they will! Ensure you integrate your revision with the school programme.

- Try and complete some revision work with other students in your group.

- Start with topics you know.

- Learn your examples and case studies (know MEDCs and LEDCs).

- Use marked work and class notes and practise past questions.

- Read all those articles, magazines snip-its or newspaper cuttings you've collected and been given .

- Use post-it-notes and highlighters, and background music, if it helps you.

How to boost your grade

Work through practice questions

This book is designed to help you get better results.

- Look at the grade A and C candidates' answers, see if you understand where they've done well or where they have slipped up. Make sure you understand why the answers given are correct.

- Try the exam practice questions and then look at the answers.

- When you feel ready, try the AS and A2 mock exam papers.

If you perform well on the questions in this book you should do well in the examination. Remember that success in examinations is about hard work, not luck.

Plan and time your answers carefully

- Spend the first few minutes of any assessment reading through the whole question paper.

- Answer the question you think you can do best first of all.

- Don't write out the question, it wastes time and space.

- Use the mark allocation (1minute = 1mark) and number of lines, to guide you on how much to write.

- Plan your answers; do not write down the first thing that comes onto your head. This is absolutely necessary for extended prose and essay questions. Remember it is better to write three average essays, or pieces of extended writing, than one excellent one and two poor ones.

- Allow time to read through your answers and remember any answer is better than no answer.

Follow the ten tips below

1. When revising make topic summaries. From a central core theme or title add spider fashion sub-themes use these in your revision.

2. Learn to focus on essentials in your revision. Use a small selection of books, your notes and articles you've been given through AS and A2.

3. If there is a choice of questions to be made, make a choice that is right for you. Questions frequently change course and may become more difficult. In short, read questions thoroughly, they may start out easy and then become too hard or obscure for you to finish.

4. If you write a plan, use it! You are wasting your time if you don't.

5. It is important to learn your definitions. Commit time to learning the "vocabulary" of the subject; you can't afford to miss out on these easy marks in the exam. They are always there for the taking.

6. Make it easy for the examiner. You'll be surprised how many candidates lose marks because their writing becomes illegible and untidy in the examination "gallop".

7. The best geographical answers introduce geographical ideas and concepts that are then backed up by facts and figures. If you are asked for numbers or data ensure you give/state the units i.e. °C, tonnes, Km^2).

8. Remember marks don't transfer from one question to another, don't spend too much time on any one question. In the same vein, giving four reasons for an answer when two are required wastes time.

9. You can't write a physical geography essay without drawing some simple sketch-maps or diagrams. You can invariably draw diagrams in your human essays too.

10. Always attempt to reach the highest level; remember all Boards use levels marking. The highest level invariably expects you to broaden a theme (especially in synoptic assessments) possibly using examples, commenting on an exception, an anomaly or trend. And, of course by introducing understanding from various areas of the specification.

This book is designed to help you gain the grade you deserve. Be positive, work hard and good luck.

Questions with model answers

C grade candidate – mark scored 10/15

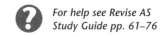

For help see Revise AS Study Guide pp. 61–76

Examiner's Commentary

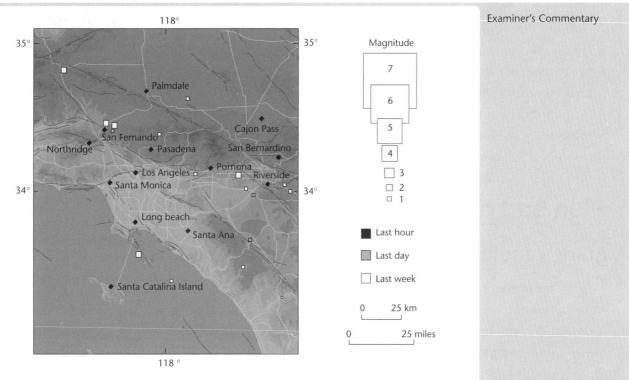

Source: US Geological Survey

(1)   Study the map, above, which shows recent earthquake activity in California and Nevada.

   (a) Describe the impacts of such events in this area.   **[4]**

   (b) What impact do earthquakes have in LEDCs?   **[4]**

   (c) 'If you understand how plate tectonics operates, then you can manage earthquakes appropriately.' Explain this statement for an area like that shown on the map.   **[7]**

(a) The biggest impacts would be seen near to the epicentre of these earthquakes. Much infrastructural damage ✔ *would be seen in and around this area. Fires and damaged buildings would catch light and be rased to the ground, possibly due to the effects of liquefaction* ✔ *rupturing pipes. Away from the epicentre, in MEDCs, very little damage occurs* ✔ *.*

(b) In an LEDC, buildings will more than likely collapse near the epicentre as building regulations are often ignored or are less rigid. There would be great infrastructural ✔ *damage, with many fires and collapsed roads. Communication links and water pipes, gas pipes etc, essential for rescue are likely to be destroyed or damaged. Liquefaction on the* ✔ *coastline and damage would have spread further than in an MEDC. Looting and crime would possibly be more of a problem in an LEDC, such as Turkey 1999* ✔ *.*

(c) The San Andreas fault in California is known to be a conservative margin in which plates slide past one another. This allows scientists to monitor plate

> *Reasonable, some use of appropriate terminology.*

> *Able to link processes to events – less good on relating ideas to actual LEDCs.*

Questions with model answers

C grade candidate continued

 For help see Revise AS Study Guide pp. 61–76

Examiner's Commentary

movement more effectively as they know in which direction the plates are moving. The fault however is not always clear ✔ and a lack of understanding hinders plate monitoring and the giving of early warnings. If scientists know what type of plate boundary it is then they may well be able to predict its likely effects and as a result plan effective rescue schemes and evacuation routes ✔. In understanding plate tectonics and their movements, people are able to see what areas are most likely to be affected and hopefully educate people of the hazards of the area (e.g. Kobe ✔ - following the 1995 earthquake) scientific knowledge along with capital have helped monitoring greatly. Early prediction in the future will help many ✔.

This answer weaves a rather tortuous path through plate tectonics.

Doesn't focus on the resource offered. Kobe is not in California!

Overall, (c) needs a greater emphasis on an exemplified and clear strategy.

A grade candidate – mark scored 12/15

(a) At the epicentre of earthquakes in this area is where the force is strongest. This would be where most buildings collapsed, and perhaps where most lives are lost. However, in the built-up area of Los Angeles ✔, where there is a much higher population density, and building density, more buildings and roads would be damaged. The presence of gas and electricity pipes would cause fires, and there may be more casualties here due to the overcrowding of the land. In areas further from the epicentre ✔, such as Nevada, damage would be minimal and people would only feel a short, small shock ✔.

Talks about relative impact.

(b) When earthquakes occur in LEDCs, where there is less money available to the government, warning systems would not have been as good, so damage caused would be greater ✔. It is likely that the buildings would not be built to withstand earthquakes of this size - in shanty towns ✔ house will fall down. Buildings as well as infrastructure would be destroyed, so rescue services would be less effective and the earthquake would have a worse effect, killing more people ✔.

(c) Scientists have long studied plate tectonics, especially in areas such as California, where they have the problem of the very active San Andreas fault. By studying tectonic activity and how the earth changes before major earthquakes, they are able to predict a major earthquake ✔✔ when they see similar activities, for instance minor shocks become more frequent before an earthquake of high magnitude. This knowledge can be used to create hazard maps ✔, which locate areas which are most at risk from an earthquake. In these areas such as San Francisco and Los Angeles, they are then able to modify buildings to withstand pressure exerted by an earthquake ✔, and also bridges and roads, such as the Golden Gate Bridge in San Francisco. In this way, scientists help to increase knowledge, so that preparedness increases and damage ✔✔ is kept to a minimum.

This student clearly understands plate tectonics and has studied pre-earthquake signs. Not much on theory.

This point really hones the focus of the question.

Part (c) well written and purposeful – several management strategies that are clearly supported and balance the answer.

C grade candidate – mark scored 11/15

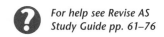

For help see Revise AS Study Guide pp. 61–76

Extended writing

Examiner's Commentary

(2) Outline the nature, causes and consequences of geological/geomorphological hazards. **[15]**

Volcanoes pose a geological hazard to any people buildings/property situated nearby. Volcanoes are caused by a build-up of heat and pressure deep within the mantle of the earth ✔. As this pressure becomes greater the hot magma/molten rock rises towards the surface of the earth and is released at the surface, often in an explosive manner via a volcano ✔✔. The pressure and heat build-up beneath a volcano is wholly attributed to the subduction of a dense oceanic plate below a lighter continental plate. In this way the oceanic ✔ plate is turned to molten rock later erupted by a volcano if the pressure becomes too great within it ✔.

Good on nature and causes. Less good on consequences.

No case study or example material.

A volcano poses physical hazard to people living nearby through its eruptions of lava, falling tephra rocks, thicker ash clouds and boiling gas causes nuee ardentee. These however are only the primary mechanisms through which a volcano can be a hazard ✔. Subsequent rainfall can initiate lahars. Sulphur dioxide gas released into the atmosphere can form acid rain and land is rendered infertile. Volcanic eruptions also contaminate water supplies and lead to disease and famine ✔ throughout populations nearby.

Landslides can be a hazard to people living on unstable land or below it. They are caused by deforestation of slopes (the trees bind the slopes together) and by excess human habitation on steep contours of hills etc. ✔

Brief sojourn into landslipping. Planning is poor in this example and detracts from the range of ideas being offered.

Often initiated by a heavy rainfall the land becomes soaked and can fall away from the hillside to which it is attached, thus posing a hazard to all it encounters as it moves down the slope at anything up to 100 km/hr.

Earthquakes can pose a hazard to people, less so through their primary effect of ground shaking but more via their secondary effects, which lead to building collapses and falling debris. Earthquakes are caused by the movement of ✔ continental and oceanic plates against one another resulting in sudden movements of one or the other as stresses and tension are released. They are commonly found above destructive plate boundaries above the Benioff zone of the ✔ subducting plate. The country of Japan is situated above a destructive plate boundary and as such is likely to be subject to a diverse earthquake in the future.

Avalanches are caused by the build-up of snow on steep slopes, as the weight of snow or snow increases segments of compacted snow break of the mountain side and move downhill at terrific speeds devastating everything in its path. Apart from the huge volume of snow and ice cascading down the slope much damage is done by the resulting pocket of air pushed at the head of the avalanche ✔.

Remembers, suddenly, there are hazards other than those with a volcanic or earthquake link!

Questions with model answers

A grade candidate – mark scored 13/15

 For help see Revise AS Study Guide pp. 61–76

Examiner's Commentary

Geological hazards can be very varied but the two most recognised hazards are earthquakes and volcanoes. Both of these hazards have been responsible for major disasters in recent years.

> *Weaker candidates might have begun with a list of hazards.*

An earthquake is caused by the movement of two plates together and occurs when energy builds up along a plate ✔ boundary and then the plates slip thus resulting in a sudden release of energy which dissipates as shock waves. It is this release of energy ✔, which causes the damage attributed to earthquakes. For humans the hazards caused by an earthquake can vary greatly with the strength of an earthquake and the proximity of the epicentre and depth of the focus of the earthquake ✔. In a strong earthquake or where building quality is poor a major hazard will collapse buildings, because of the shaking. In Agadir in 1960, a weak earthquake, 5.8 on the Richter scale caused massive destruction and the death ✔ of 12,000 people. This was because of a high population density and poor building construction, which was not designed or capable of withstanding an earthquake. Many buildings collapsed trapping and killing those inside ✔. Another earthquake hazard is the subsidence of buildings and liquefaction. Both of which can cause the collapse of buildings. In an earthquake hazards such as collapsed electricity lines and explosions from ruptured gas pipes also occur ✔.

> *Demonstrates breadth of knowledge related to some good case study work.*

For a volcano the hazards depend greatly on the type of its magma. Lava from the volcano is usually a minor hazard ✔✔ because it is slow moving and lava flows can often be predicted. Lava flows are extremely destructive to houses and to vegetation. The greater hazard for volcanoes tends to come from pyroclastic flows which involve clouds of hot rocks ✔ and ash racing at high speeds down a volcano and destroying and killing everything in the way. Pyroclastic flows or nuee ardentee were responsible for the destruction of the city of St. Pierre in Martinique. Falling rocks thrown ✔ from the volcano and ash could also present a hazard - clouds of ash can often destroy crops and cause famine ✔. Volcanoes erupt because of a build upon pressure within their core as magma and gases rise up from the mantle. As the pressure grows minor eruptions may occur until the pressure reaches a critical point when a major eruption may occur.

> *Clear understanding of the nature of the processes involved, some weaknesses on causes.*

> *Fort de France is the capital city of Martinique.*

Both earthquakes and volcanoes can cause secondary hazards such as landslides and avalanches. In California because of tectonic activity ✔ 95% of slopes are unstable. Volcanic eruptions can force water out of the crater, which will cause mudslides like those, which wiped out villages around Nevada del Ruiz in Colombia. Gases from volcanoes can be toxic and will kill people ✔ and wildlife if they escape.

> *Concentrating on earthquakes and volcanoes can enable a top mark to be gained.*

Geological hazards pose a hazard to life and also property. They can cause great economic damage and social damage. All of these can be caused by the range of geological hazards.

Exam practice question

Structured question

(1) (a) (i) Draw a simple diagram of a rift valley; label it clearly to show fault scarps, blocks and faults. **[3]**

    (ii) Explain how a rift valley is thought to form. **[2]**

(b) Describe, with reference to a named example, the effects of a volcano on the lives of the local people. **[4]**

(c) The following is a description of an important rock:

'This rock, though now at the surface, was formed at great depths below the surface of the earth. It contains large crystals which formed as the original molten rock cooled slowly and solidified.'

    (i) Name the rock. **[1]**

    (ii) Name an area consisting of this rock and describe the main features of its landscape. **[5]**

    (iii) State one reason why such areas are usually sparsely populated. **[2]**

(d) The sketch section shows a distinctive rock structure.

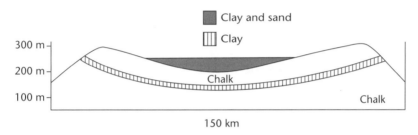

    (i) Name this structure. **[1]**

    (ii) Mark onto the section a scarp slope and the water table. **[3]**

[AQA (SEG)]

Answer

(1) (a) (i) Fault scarps, blocks and faults need to be added.

    (ii) Two or more parallel faults and subsidence of the area between the faults.

> **Examiner's tip**
>
> *Many of the specifications have returned at this level to requesting basic structure knowledge. This is invariably tested with such 'fill-in' diagrams. Sounds easy, but if you haven't got the 'language of the subject', you can lose simple/easy marks.*

(b) Death; destruction; services lost; tourism; mining – for precious metals, etc; geothermal heat; rich soil; ignimbrite for building, etc. Don't forget named examples.

(c) (i) Granite or gabbro.

    (ii) Dartmoor – rounded, smooth features; tors; impermeable, waterlogged, wide-open valleys.

> **Examiner's tip**
>
> *This question asks you to relate what you see on the surface (landforms) to the nature of the underlying bedrock.*

    (iii) Thin soil, poorly drained, exposed, farming difficult.

(d) (i) Syncline, downfold, basin.

    (ii) Accurately drawn onto the diagram.

Questions with model answers

C grade candidate – mark scored 8/15

 For help see Revise AS Study Guide pp. 78–89

Examiner's Commentary

(1) (a) (i) What is climatic climax vegetation?   **[3]**

   (ii) Describe one way in which climatic climax vegetation can be destroyed by a change in physical conditions.   **[2]**

   (b) With reference to the above photo, explain the evidence for human contribution to the spread of deserts.   **[3]**

   (c) For a biome that you have studied, describe some of the relationships that exist between its soils and vegetation.

                    [7]

[Adapted from AQA]

(1) (a) (i) This describes the type of vegetation that dominates in a certain area, over another type of vegetation. For example the large trees that form a canopy in tropical rain forests prevail over the smaller shrubs etc below them ✔.

(ii) Frost
   Little rain

(b) This process is known as desertification. And is caused by the overgrazing of animals ✔ which eat all plants which keep soil together, thereafter soil gets blown away. There tends to be little water and/or irrigation ✔.

(c) Coniferous woodland - covered by pine trees, which have waxy needles, thus no real intake of water ✔. Water drains quickly into the soil. Because of the waxy needles they are very hard to decompose ✔, and take a long time to rot. The combination of the needles, water, heat and fermentation processes leads to an acidic soil to be formed. Below a dark humus layer a hard pan starts to form ✔. The acidity of the soil is also controlled by the parent rock ✔✔.

Words to use here: end/complex/competitive/controlling/natural vegetation.

Avoid lists. There are lines in the actual exam – fill them with prose!

Some relationships established. There are, however, a number of inaccuracies. Candidate aware, at a basic level.

A grade candidate – mark scored 14/15

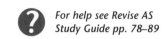 *For help see Revise AS Study Guide pp. 78–89*

(1) (a) (i) The vegetation type that finishes a succession. The vegetation type is in equilibrium with the environment ✔. The vegetation is the most dominant and is the best competitor in a vegetational succession ✔.

Answers here have to be 'definitive'. Definitions need to be learned!

(ii) If an area becomes regularly flooded by the sea due to a rise in sea levels or the isostatic sinking of the land, the increased salinity of the soil will kill off the climax vegetation ✔, which will be replaced by salt-tolerant species ✔.

(b) The desert has spread furthest into the Savannah area where there is water for animals ✔ and villages. Where there is no human intervention there is no desert spread. Humans are allowing their animals to overgraze areas; poor farming practice allows soil erosion ✔ to occur ✔.

The resource has been used and own knowledge makes a contribution.

(c) Coniferous forest – because the soil is often slightly acidic and does not have many nutrients ✔ few plant species but coniferous ✔ forest can grow on the soil. The coniferous forest does influence the soil with the thick leaf litter layer caused by the hard to decompose pine needles ✔. These contribute to the soil's acidity by producing a layer of acidic mor humus. The soil is often friable and shallow which prevents other trees from growing in the soil ✔ and means the coniferous trees grow lots of shallow roots that spread out over a large area to increase stability in the loose soil ✔. The lack of nutrients means that deciduous trees which have to replace all their leaves in one go ✔ struggle to survive while coniferous trees have a continuous shedding and growth of needles ✔.

Clear and concise answer, which displays an understanding of the 'links'.

Overall excellence displayed by the A-grade candidate.

C grade candidate – mark scored 10/15

Extended writing

(2) Describe some of the factors that could account for the changes in vegetation in an area over time.

Be prepared for 'change/link'-type questions for soils and ecosystems.

Succession is the name given to the change of vegetation over time ✔.

With no human activity an area will follow a progression of natural succession, until a climax community is achieved. The process starts on the origin rock of an area, microbes decay the rock depositing it and themselves when they die. As new more evolved species become able to live in the area it will increase that areas fertility and nutrient content, making it habitable for large/more sophisticated and substantial species. This succession takes place in all birds, insects, animals and plants until the climax community is established ✔ ✔. Climax communities include tropical rainforest in equatorial regions and temperate deciduous forest in the UK etc ✔.

Succession correctly identified.

Questions with model answers

C grade candidate continued

 For help see Revise AS Study Guide pp. 78–89

Examiner's Commentary

If human activity interrupts and succession destroys the climax community the secondary succession will follow, returning a secondary climax community ✔. Natural interruptions occur to cause the plagio stage, and promoting the secondary succession.

Attempts to sequence changes but runs out of ideas/knowledge.

An example of changing vegetation of an area can be seen in the distribution of the tropical rainforest. If deforestation is done at a small-scale subsistence level such as the slash and burn type of farming practised by indigenous tribes in Brazil then the tropical rainforest can be used sustainably and once left, since the area is relatively small, secondary succession can occur and quickly restore the forest back to its original climax state ✔.

Use of 'continental' examples works less well here.

Other human factors causing the change in vegetation over time can be seen in the Sahel desert. Desertification a process by which continuous erosion of soil (Sub-Sahara) causes it to become infertile.

Monoculture, excessive use of pesticides, overgrazing, lack of protection from the elements, causes nutrients in the soil to be depleted rendering them infertile ✔. When the soil is infertile little can grow and bind the soil; therefore it is blown or washed away ✔.

Farming and urbanisation change the vegetation in an area, indefinitely, where no surrounding succession can occur.

A plan might have helped this candidate.

Natural factors include landslides, mudslides, glaciers, climatic factors - flooding/drought all cause succession and climax states to be suspended ✔.

If humans were to be extinct the world's vegetation would eventually return through succession to its natural state of climax ✔. Man has to be the most important factor in accounting for the change over time in the vegetation of an area.

Overall: some inconsistencies and lack of balance. Needs greater organisation.

A grade candidate – mark scored 13/15

The changes that occur overtime in the vegetation of an area are known as succession ✔. A good example of where this can be seen is Holkham Beach in Norfolk, on the sand dunes. On my new patch of land a new species will appear and grow, on the embryo and fore dunes at Holkham sea rocket and marram grass (both resistant to salty and dry conditions) appear these are the pioneer species ✔.

Candidate roots answer successfully in a suitable case study!

A grade candidate continued

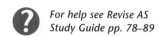
For help see Revise AS Study Guide pp. 78–89

Examiner's Commentary

These pioneer plants have the effect of stabilising the sand with their roots etc. This allows water to be trapped along with nutrients. Gradually conditions improve enough for new species to colonise the area, which perhaps would not survive here before. This succession continues with each new species that comes along replacing another as conditions continue to improve ✔. Looking at the dunes at Holkham, the succession of different species can easily be observed by taking a transect back from the embryo dunes to the mature dunes and slacks ✔. The further you head back the more growing conditions improve and diversity of species can be seen. Different species add new ✔ nutrients to the soil, and eventually certain wildlife can also be found to colonise the area, such as rabbits that also add nutrients to the soil through their faeces, as they survive on the existing vegetation.

Answer uses the 'language' of the topic.

Eventually there becomes a point when succession begins to cease and the vegetation stabilises to equilibrium with the environment. The final type of vegetation is known as the climax vegetation, and can be dependent upon many things ✔. Many geographers believe that the principal factor in determining an area's climax vegetation is the climate. For this reason the vegetation would be known a climatic climax vegetation.

An easy read – it flows!

Unusually in Norfolk however the climax vegetation at Holkham is an area of extensive pine forest. This does not fit the idea that it is entirely dependent on climate, as coniferous trees are usually found in colder environments with very high rainfall, such as the North Western side of the UK. The reasons for this climax vegetation must therefore be put down to other things ✔.

Other factors that exist are things like geology and human impact ✔. At Holkham the geology may be the reason, and if so the climax vegetation would be known as geo-climax vegetation. The area's geology is such that it would provide the bases for the acid soils, which would suit the growth of conifers ✔.

Humans are very likely to have an impact in the vegetation type at Holkham. This would result in a plagioclimax vegetation ✔. Holkham is a very heavily managed area, controlled by English Nature, as the sand dunes are literally being worn away. For this reason many paths have been installed in the dunes and trees planted ✔. It is at this stage that one might reasonably assume that the pine trees had in fact been planted, and are in fact a plagioclimax ✔ community.

Maintains a 'local', small-scale focus.

Judgemental – but fine!

Holkham just gives one example of how vegetation can change over time, and the many reasons that can account for it. Other examples would easily be seen in the ecosystems of lakes and perhaps mountains ✔.

Overall: competent, organised, reasonably accurate and well supported. Consistently relevant to the theme.

Exam practice questions

Structured question

A *Answers on p. 17*

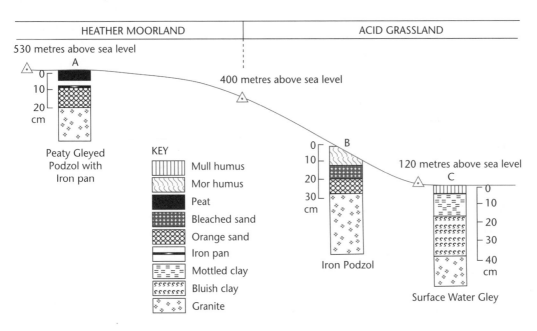

(1) The diagram above shows a soil catena that has developed on the slopes of Bennachie Hill, Aberdeenshire. The parent material is granite.

(a) (i) What is meant by the term parent material? **[1]**

(ii) State what is meant by a soil catena. **[2]**

(b) (i) Suggest why there is a peat horizon in soil A. **[2]**

(ii) Suggest why there is a bleached horizon in soil B. **[2]**

(iii) Suggest why there are mottled grey and blue-grey horizons in soil C.

[2]

(c) The natural vegetation on Bennachie Hill may be described as a plagioclimax.

(i) What is meant by the term plagioclimax? **[1]**

(ii) Suggest how this plagioclimax may have developed. **[3]**

(iii) Describe and explain how subsequent modifications of vegetation by humans, in areas such as Bennachie Hill, produce a variety of plagioclimaxes depending on local conditions. **[4]**

[Adapted from UCLES/OCR]

Extended writing

(2) Explain briefly how man has interrupted and disturbed the nutrient cycle. **[25]**

Answers

(1) **(a)** **(i)** This is the bedrock, i.e. the principal source of weathered rocks from which the soil has formed.

(ii) Soils vary according to the gradient and position they occupy on a hillslope. In order to describe lateral variation of soils, the term soil catena is used. A repeated sequence of soil profiles.

Examiner's tip

A straightforward recall question. Basis of this topic, i.e. 'how do soils form'. Plan what you want to say carefully and accurately.

(b) **(i)** The peat horizon is likely to be a blanket bog – associated with human activity, the soil becomes increasingly wet as a result of people reducing evapotranspiration losses from the soil by removing the original upland treecover. As these areas receive high rainfall, the now treeless landscape became saturated and peat formed.

(ii) This layer is strongly eluviated by organic-metal compounds called chelating agents. These combine with metallic ions (Fe, Al), which readily pass down through the profile. This leaves an extremely bleached/ashen-coloured Ea-horizon.

(iii) Soils that are waterlogged restrict the penetration of oxygen into the profile. Mottling occurs when local pockets of air re-oxidise compounds in the soil. This happens in soil pores, cracks and live root channels.

Examiner's tip

Catenal sequence questions enable examiners to test all areas of the topic – processes and effect in particular! By picturing the profile in your mind, of a typical cross-section of soil being examined it should be easy to deal with the quirks imposed by the slope! In the question you are asked to outline specifics on how man affects a small biotic community; you must know these as well as the old faithfuls of tropical rainforests and the prairies.

(c) **(i)** Where natural vegetation is subject to a great deal of human interference.

(ii) The clearance of upland forests has allowed heather to take advantage of newly created, open, exposed and often nutrient-deficient sites. Heavy grazing and frequent burning prevents the regeneration of forest. Heather maintains its cover.

(iii) Poor drainage causes blanket bog; overgrazing can maintain heathland; grazing and burning stops and oak or pine forest returns; liming and fertilisation can help improved pasture to start to appear.

(2) You might choose from some of the following:
- Removal of vegetation by man – for crops.
- Removal of ground cover increases run-off – leaching.
- Changing farming practice – rotation/fertilisers/eutrophication.
- Dumping sewage at sea.
- Pollution.
- Over-exploitation.
  The best answers may well range into sustainable development.

Examiner's tip

This extended piece of writing asks simply for you to comment on how man interferes with the nutrient cycle; it's not concerned with how it adapts, neither does it ask for examples. The best answers will comment on how man can influence the sustainability of ecosystems. Clearly, some well-drawn and appropriately placed diagrams will really benefit your answer.

Question with model answers

C grade candidate – mark scored 9/15

 For help see Revise AS Study Guide pp. 21–33

Examiner's Commentary

The hydrographs below show the effect of heavy rainfall received in two catchments in Norfolk during November 2000, and a similarly sized catchment in Derbyshire.

Hydrograph questions are common. Know all there is to know about them. For the examiner a single hydrograph question conveniently tests many hydrological relationships.

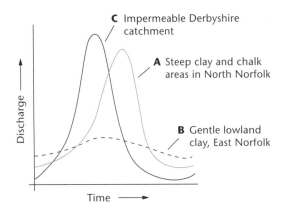

(a) 'Two very different catchments in Norfolk contributed to the hydrograph shapes shown above.' Describe and explain, briefly, the differences between the two.   **[4]**

(b) 'The sandstone found in parts of Derbyshire is impermeable.' How would you account for the shape of the hydrograph?   **[4]**

(c) 'Erosion processes dominate during periods of high flow. Depositional features emerge and appear in channels during periods of lower flow.' For a named river, describe where most depositional features appear and explain why deposition is dominant in the part of the river used as an example.   **[7]**

(a) Catchment A has a sudden increase in discharge and then a fairly sudden decrease, this is due to catchment A having steep valley sides ✔ so run-off will be very quick due to gravity and there will be not much water sinking into the soil so there is little through flow ✔. Surface run-off is quick hence the sudden increase and sudden increase and sudden decrease. Catchment B is a more gradual increase in discharge and decrease, this is due to the gentle valley sides allowing water to be absorbed into the ground and then through flow can occur ✔, which is slower than surface run off hence the slight further delay before the discharge showed a notable change.

Comment on peak, lag-time, run-off, steepness of slopes and percolation.

Descriptive.

Explanation is incomplete here.

(b) The rock is impermeable to water so therefore no through flow can occur so the water has to undergo surface run-off which is a faster process than through flow ✔ and so the discharge will have a sooner increase and it will be very sudden, but it will also decrease suddenly a surface run-off does not tail off slowly it stops quickly when there is no more water on the surface ✔.

C grade candidate continued

For help see Revise AS
Study Guide pp. 21–33

(c) E.g. The River Dane, Derbyshire. On a fairly flat plain across which the river flows there is a braided channel ✔ *where the river has been widened through horizontal erosion instead of vertical erosion. But the channel has become braided due to the shallowness of the water and its slow movement* ✔ *due to the gradient so deposition has occurred leaving a variety of smaller channels running through small islands of deposited sediment. Its deposits appear because the water doesn't have enough energy to transport the material* ✔. *This depositional feature dominates due to the gradient being flat and small volumes of water* ✔.

Examiner's Commentary

Channel features only. Limited lack of locational knowledge.

A grade candidate – mark scored 14/15

(a) The hydrograph for catchment A is steeper with a higher peak discharge than hydrograph B ✔. *This is due to the steepness of the valley sides. As the valley sides are steep* ✔, *the water will rush into the river system much quicker* ✔ *and will therefore reach the discharge point marking more quickly. This not only gives a steeper rising limb but also a higher peak discharge. The falling limb falls more quickly because all the water is through the system quicker. B is shallower because it takes more time for the water to pass through* ✔.

Higher order description and understanding.

(b) C starts off with a lower discharge to B as there is less groundwater flowing through the impermeable system ✔. *The rising limb is steep and the discharge is high due to the fact that no water seeps into the system* ✔, *it all runs off quickly into the river channel* ✔. *The volume of water passing through is therefore very high. The discharge is low after the storm because there is again very little water flowing through the system due to the impermeable rock* ✔ *and the fact that all of the water has passed through.*

Any discussion of flood hydrographs requires you to understand the factors that affect the timing and volume of river discharge.

(c) E.g. River Rheidol, West Wales. In the middle reaches of the Rheidol ✔, *mixed in with erosional features, there are many depositional landforms. On the inside of many meanders there are massive slip-off slopes caused by the deposition of material on the slow moving insides of bends* ✔. *These slip-off slopes are so large because of times of flooding. Also in these reaches, where the gradient is suddenly decreased to give the river less energy to hold the load* ✔, *there are braided channels caused by the deposition of material in the middle of the channel* ✔. *Near the mouth of the river, at the harbour, in Aberystwyth, the river is very slow moving due to the shallow gradient coupled with the effects of the tide. This has caused silting up of the river and harbour bed* ✔, *this occasionally needs to be dredged to prevent the river from re-diverting its course. A weir has been installed to prevent this but it has not done its job very efficiently* ✔.

Located and uses local knowledge and has a good range of features for the river chosen!

General commentary:
- *For hydrograph questions be prepared for comparison and measurement approaches.*
- *Know and understand how basin shape and size, relief, geology, soils and vegetation and how humans influence aspects of basic hydrology.*
- *Be able to comment on hydrological relationships using examples.*

Exam practice question

Structured question

A *Answer on p. 21*

(1) **(a)** The diagram below shows the long profile of a river in the UK.

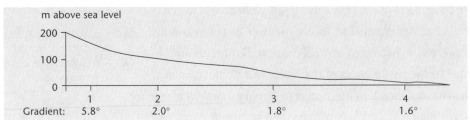

(i) Suggest two reasons that might explain why the profile of the river is ungraded at present. **[2]**

(ii) On the diagram above draw a graded profile. **[2]**

(iii) What does the term 'graded' mean? **[3]**

(b) The graph below relates average velocity in the stream channel shown above to distance from its source.

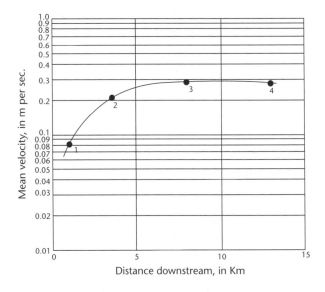

(i) Describe the relationship between velocity and gradient in the above stream. **[4]**

(ii) What three factors influence stream velocity? **[3]**

(c) Sand, silt and clay are three size classifications of sediment.

(i) Which class of sediment requires the highest velocity to entrain it (pick it up)?

Explain your answer. **[3]**

(ii) Which class of sediment requires the highest velocity for it to be transported?

Explain your answer. **[3]**

[Adapted from Edexcel (ULEAC)]

Answer

(1) (a) (i) Reasons for the ungraded nature include:
- The presence of rapids and waterfalls due to rejuvenation.

Examiner's tip

Straightforward knowledge and understanding needed here to define the ungraded profile.

- Structural features, such as different rock types and faulting.

Examiner's tip

The concept of gradation in rivers is at the heart of an understanding of the concept of dynamic equilibrium. To understand fluvial geomorphology, you have to understand dynamic equilibrium and how it links with erosion, transportation and deposition.

(ii) Draw a smooth curve.

(iii) The graded profile represents a slope of transportation. The condition of 'grade' is attained when the energy of the river is used up in the movement of the water and sediment load, so none is available for erosion. The smooth, graded profile is also called the profile of equilibrium.

Examiner's tip

Finishing off diagrams or taking diagrams/graphs on a stage are common at AS as they are a quick way of ensuring you have a complete understanding of processes and how they contribute to a wider 'landscape' picture.

(b) (i) There is a relationship between slope and velocity and the distance from the river source. The number of contributing catchment areas, and therefore the numbers of tributaries, increase as you move down valley. So velocity continues to increase. Your answer could relate to points 1–4 on the graph. Why is there a slight dip at point 4? Explain.

Examiner's tip

Such graphical relationships are frequently examined. Offer both descriptive and explanatory comments.

(ii) Stream efficiency related to shape (width and depth, hydraulic radius); Frictional drag (Mannings equation/calculation); Gradient; Numbers of tributaries contributing to the flow (volume of water); Sinuosity; Uneven streambeds = turbulence = loss of energy and therefore speed; Flood conditions; Having to transport material slows rivers down.

Examiner's tip

Factors such as these need to be learned and remembered for the range of variables dealt with in physical geography. Many factors influence stream velocity/flow. It is important you know all of them and can link them to the range and variety of processes that dominate in the fluvial system. Be prepared for these straightforward questions. A basic knowledge of this type is essential.

(c) (i) Sand is entrained at high velocities. This is because drag has to be overcome. When this happens, the critical tractive force is said to have been reached. Coarser fragments are always the last to be picked up.

(ii) It also follows that the biggest sediments (sand) need higher velocities – competent velocities, as they are known – to transport them. Study the Hjölstrum Theory p. 23 AS Study Guide.

Question with model answers

C grade candidate – mark scored 9/15

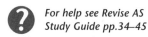 *For help see Revise AS Study Guide pp.34–45*

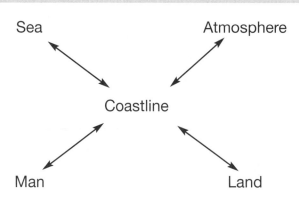

Sea   Atmosphere

Coastline

Man   Land

Examiner's Commentary

(1)   Study the diagram above.

   (a) Suggest why coasts are dynamic environments.   **[3]**

   (b) Explain how, despite being dynamic, coastlines can become irregularly shaped.   **[5]**

   (c) How does man manage coastal retreat along the coastline?   **[7]** Comment also on any disadvantages and advantages.

(a) Coasts are affected by a range of measures that come from the atmosphere, the sea (hydrosphere) and the lithosphere (land). In this way coasts develop under numerous conditions that range from natural activities such as waves which occur due to wind energy to human activities such as coastal walking and sailing ✔. All the factors that influence the coast are changing such as insolation which changes in relation to the position of the sun and cloud cover and as a result the coastal environments alter and develop in accordance with these changing conditions.

Really, only the first four lines deal with the question; the answer rather misses the point in the second half.

(b) One reason is that there could be two separate strips of rock on the coastline. One being hard (headland) the other soft rock (a bay ✔). The soft rock is eroded leaving the hard rock - the coastline retreats into the softer rock ✔. There may be sea walls, halting the retreat of the cliff. In other areas there may be no sea walls or they may be damaged, causing retreat.

To do with wave refraction – goes some way towards answering the question.

(c) One scheme is to build sea walls. These protect the cliffs by diffusing the wave energy onto the wall and not the cliff ✔. These are advantageous as they are initially fairly cheap and easy to construct and they do not need repairing often ✔. However, when they do need repairing they are expensive and are also an eyesore. Another scheme is the planting of marram grass ✔ to encourage dune development. The accumulation of sand in dunes means wave energy can spread over a great area ✔ - diffusing energy and causing little erosion. These are advantageous as they arenatural and easy to prepare ✔ but they also need a lot of maintenance ✔ and a scheme to prevent walkers from depleting them.

The candidate sticks to two schemes of management, the question doesn't ask for more! Management ideas are rather bland.

It is acceptable to practise your answers, even to plan for them! Use the left-hand margin.

A grade candidate – mark scored 12/15

For help see Revise AS Study Guide pp. 34–45

Examiner's Commentary

(a) This is the case because there are so many factors now, which can alter coastal processes and therefore change the coastline dramatically. For instance, through coastal sea defences, man has changed the natural processes of a specific area ✔; only to increase it in another ✔.

> The first sentence rather wastes precious space.

Through changes in the atmosphere, sea levels can rise and fall etc and the wind can change the direction of LSD ✔.

> Rather weakly expressed towards the end.

(b) Wave refraction focuses the waves onto the headlands causing an uneven erosional pattern ✔. Irregularly arranged hard and soft rock can have a similar effect on the coastline ✔. Man's attempt to "hold the line" can also lead to an irregular pattern to the coastline.

> Offers two points (a wave refraction point and a lithology point); neither are taken on fully.

(c) There are many strategies for coastal management. Most often groynes, wooden constructions, mainly in the breaker zone are put on the beach. They build up the beach locally ✔, but sediment deficits downdrift cause the beach to shrink and erosion to increase ✔. Artificial reefs may be constructed around a particularly bad cliff line, but this too can have bad effects on the coastline, causing embayments and build-ups of sand behind reefs ✔. Beach replenishment is another option, when sand is dredged from the sea. Sand being the best energy absorber available ✔. Adding sand can upset the dynamic equilibrium of the coastline. In fact the Government has decided that it is best not to use these schemes ✔, it is best to let the sea do its worst and then to compensate landowners ✔. The effect of these schemes is to protect those areas that are economically important and to let the sea take those areas of little worth ✔.

> This candidate clearly understands the question, and in the space available selects strategies and shows some insight into cost–benefit analysis.

Exam practice question

Structured question

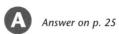

Answer on p. 25

Source: Nick Gregory, APEX Photo Agency

(1) (a) Label the photo above to show the different parts of this coastal slump, and how and why this stretch of the Dorset coastline, at Charmouth, collapsed in December 2000. **[5]**

(b) Label the diagrams below to explain how landslides might result from the interaction of physical and human factors. **[7]**

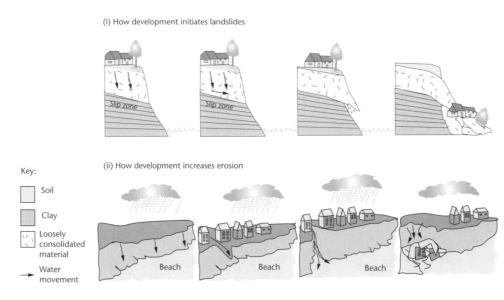

(i) How development initiates landslides

Key:
- ☐ Soil
- ☐ Clay
- ☐ Loosely consolidated material
- → Water movement

(ii) How development increases erosion

Extended writing

(c) With reference to a stretch of coastline you have studied:

(i) Describe the physical factors which have promoted landslipping or coastal erosion.

(ii) How successful were the management strategies used to combat the impact of the processes of landslipping and erosion? **[19]**

[**(b)** and **(c)** from Edexcel]

Answer

(1) (a) Labels: rainfall/waterlogged clay/steepness of cliff face/narrow beach/rotational slide/slip face/curved rupture zone/run-out. For how and why see parts of **(b)** below.

Examiner's tip

It's AS – expect these simple labelling exercises on pre-drawn diagrams or photos. They test your vocabulary and your process knowledge in a practical way. Do use accurately placed arrows!

(b)(i) P1. Increased loading of buildings/vibration from human activity/impact on the hydrological cycle/down slope dip.

P2. Effect of tree roots/percolation through the sandstone/lubrication surface on the clay.

P3. Waves create instability at the foot of the cliff.

P4. House falls from the cliff.

Examiner's tip

This question takes the theme forward, increasing both the complexity and the rigour of the question, as structured questions should. Ensure that you cover both areas – those of landslipping and erosion – in your answer.

(ii) P1. Run-off spread along the cliff/bare rock is susceptible to sub-aerial erosion/storms = overland flow.

P2. Roads increase permeability/concentrated run-off into drains/houses add to the loading.

P3. Localised gulleying/natural beach/no defences.

P4. Gully washes out and the cliff collapses.

(c) (i)

	Landslipping	Erosion
Geological factors	Soft rock/dip of rocks/ bedding planes.	Soft strata/weak lithology, as left.
Geomorphic factors	Steepness of cliff face/ width of beach/nature and strength of wave action.	Fetch/wave types/Long shore drift/clapotis, etc.
Meteorological	Heavy rainfall/freeze/thaw/ snowfall.	Storms/aspect of coast/winds.
Biogeographical	Biological weathering/ vegetation holds back earth movement.	Easy on unvegetated cliffs.

Factors chosen should relate to the area chosen.

(ii) Landslipping: drain/sheeting/vegetate/infill any gullies/change the angle of the cliff/toe protection/nourish/rip-rap/revetments/groynes.

Erosion: nourish/armour/off-shore strategies.

Examiner's tip

Only the strategies are offered here as the success depends on the location/examples chosen. Some idea of Cost Benefit Analysis is necessary, as is an understanding of current governmental strategy along the coastline. A good synthesising answer will be necessary here.

Question with model answers

C grade candidate – mark scored 11/15

? *For help see Revise AS Study Guide pp. 46–60*

(a) What are greenhouse gases? [3]

(b) Describe the source and reasons for increasing amounts of greenhouse gases in the atmosphere. [5]

(c) Why should we be concerned that greenhouse gases are increasing? [7]

Examiner's Commentary

(a) These are gases that contribute to the greenhouse effect. The greenhouse effect is an effect whereby the earth is becoming increasingly like a greenhouse in that heat etc is allowed to enter the atmosphere but not escape heating the earth up ✔. These gases stop heat etc getting out, contribute to the greenhouse effect, and so are termed greenhouse gases ✔.

Poor response, as it digresses into talking about increased 'greenhouse' effect!

(b) CFCs have increased (or are predicted to increase massively) as a result of a massive increase in consumer goods which contain them and release them into the atmosphere ✔. The other problem is their increasing availability in the Third World/developing/NIC countries ✔. Products such as aerosols and refrigerators are becoming more and more widespread across the world and expected to continue to do so ✔. In China there are massive fears about the amounts of waste refrigerators and the CFCs they release ✔.

This answer is good, as it ranges into the wider 'development' issue. Does digress into other gases too.

(c) The damage that they and the greenhouse effect will cause to the environment and the microclimate by increasing the temperature in the world's atmosphere could have massive consequences. Ice caps would melt, causing sea levels to rise ✔ - which could be devastating for much of low-lying Europe, Yangtze and Ganges basins in Asia, and East Coast America ✔. A rise of 50m would displace 2 billion people. This would mean a massive loss of ecological habitats that might cause deserts to spread and the tropics to expand ✔. Animals suited to colder climates would be pushed out and many habitats would be destroyed or change so as to cause mass extinctions ✔. There is also the worry that the likes of CO_2 and CFCs and other harmful pollutants would increase too ✔.

Good on human impacts and good, short examples.

Needs more on temperature, modelling and 'debate'.

A grade candidate – mark scored 13/15

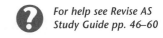

For help see Revise AS Study Guide pp. 46–60

Examiner's Commentary

(a) The gas allows sunlight of a high wavelength to pass through from the sun ✔. The sunlight hits the earth, and loses energy. When it tries to `bounce´ out of the atmosphere again, it can´t because the lower wavelength is refracted by the greenhouse gases trapping the light in the atmosphere ✔, insulating the earth, just like a greenhouse ✔.

Starts to use the language of the issue!

(b) Carbon dioxide comes from the burning of fossil fuels. The two main culprits are cars and coal power stations ✔. Despite the development of more efficient engines and the reduction in coal burning, the huge increase in the number of cars in MEDCs, but more importantly in LEDCs, will cause it to increase ✔. Even the Kyoto climatic agreement, at its most optimistic, only aimed to slow CO_2 production ✔.

Focuses on the CO_2 theme reasonably effectively – knows sources and wider issues related to the problem.

(c) There has for many years been a debate within the scientific community about the consequences of the release of greenhouse gases. Scientists believe that it is the resultant greenhouse effect that might be causing or at least contributing to the current raising of global temperatures ✔✔. If this is the cause then the consequences are dire. Polar ice caps will continue to melt causing global sea levels to rise ✔. Over the next 50 years low-lying areas will be submerged ✔. The melting ice caps will alter ocean currents, e.g. the Gulf Stream, affecting the weather patterns of the UK, making it way cooler ✔✔. Global weather temperatures will change too, with wetter summers and more extreme winter weather conditions ✔. All these problems will have a great impact on both human and all other life on this planet!

Understands the 'debate', i.e. temperature increases and links to other systems.

Perceptive with regard to human impacts.

Good on consequences.

Exam practice question

Structured question

A *Answer on p. 29*

(1)

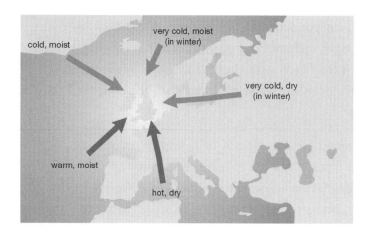

(a) (i) What are the main factors controlling the direction and speed of the wind? **[4]**

(ii) Define the term 'air mass'. **[2]**

(b) (i) Identify the air mass areas affecting the British Isles (diagram above). **[4]**

(ii) What are the atmospheric characteristics of the source region of continental polar air? **[2]**

(iii) Describe the typical weather associated with continental polar air in winter in the British Isles. **[4]**

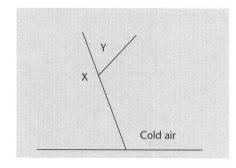

(c) The diagram, left, shows a type of front.

(i) Name the type of fronts associated with mid-latitude depressions. **[2]**

(ii) Identify the type of air found in areas x and y. **[2]**

(d) Explain the processes that produce orographic rainfall. **[10]**

Answer

(1) (a) (i) Principal ones are pressure gradient, gravity, Coriolis force and friction.

(ii) Air masses are large areas of air that have similar properties of temperature and humidity.

Examiner's tip

Process knowledge; if you don't understand the basics, you won't get very far with meteorology! Being able to define the terminology of the subject is vital at both AS and A2 levels – questions like these are common.

(b) (i) Air masses include arctic, polar maritime, tropical maritime and tropical continental.

(ii) Cold and dry.

(iii) In December to February the air is cold and there is little moisture available. Moisture may be picked up from the sea and lead to snow showers, especially in the east. Wind chill is high.

(c) (i) Cold occlusion.

(ii) x = colder air and y = warm air.

Examiner's tip

To answer this effectively, you should know and understand the sequence of weather that occurs before, during and after a front (p. 52 in the AS Study Guide). The diagram gives you minimal information – you have to be able to apply your own knowledge.

(d) The movement of air causes orographic rainfall across an area of higher relief. Precipitation increases with height. Evaporation, condensation and precipitation need to be referenced. Reference to actual reasons will probably score the higher marks. The complexities of West Coast relief and orographic or relief rainfall could be mentioned. This answer would benefit from the inclusion of a diagram.

Question with model answers

C grade candidate – mark scored 10/15

For help see Revise A2
Study Guide pp. 21–36

Examiner's Commentary

The photo on the left is of Navy Board Inlet, Bylot Island, Nunavut, Canada.

Source: Reproduced with permission of the Minister of Public Works and Government Services Canada, 2001 and courtesy of the Geological Survey of Canada.
Photo by: D. Hodgson

(a)   Outline evidence from the photo to suggest that this area might well pose a great challenge for human habitation. **[4]**

(b)   Despite the problems such northern areas create in terms of habitation, human populations are increasing in these northern marginal areas. Give two reasons why. **[4]**

(c)   In areas less extreme than Nunavut there is increasing conflict between newly arrived and indigenous populations, who have a concern for the natural environment. Outline the nature and extent of these conflicts. **[7]**

(a) To the south of the photo the land appears to be quite shallow in terms of its slope. In the northern part of the photo there are more rugged peaks, the sheer steepness of these mountains would mean that very little could be built on them because of the risk of landslides and rockfalls ✔. Much of the mountain area is covered in ice and ice caps, this would lead to an extremely cold climate and therefore difficulties with water supplies in pipes freezing up ✔. Also communications would be extremely tricky avoiding the mountains and the ice, you'd have to take many detours ✔.

Consider the style of this answer. Is it as tight and informative as one would expect for A2?

(b) - Many people move there because of work. Some for research into the area, other filming documentaries on it. Many areas have rich resources like oil, coal. These industries start up communities of workers' families ✔.

- People also want to explore the wilderness and may wish for a change in lifestyle from the everyday hassle of smoggy cities to the non-polluted and isolated mountains that clearly make up this area ✔✔.

Not enough on back to nature and isolation!!

C grade candidate continued

For help see Revise A2
Study Guide pp. 21–36

Examiner's Commentary

(c) In cold areas such as Antarctica there are conflicts between countries. With colonial rights to this country Australia owns up to 43% of Antarctica, almost the size of their own country ✔. They wish to start transforming their present research centres into hotels for tourists. Britain disagrees with this exploitation They only allow tourists to cruise the coastline, with none of them setting foot onto the land itself. Britain´s argument centres on the fact that man will be destroying another area of natural beauty, `just through greed´. In northern Siberia there are frequent confrontations between Eskimos and oil diggers ✔. The oil men wish to do their job, but are confronted with traditions and the rights of local people ✔✔.

Much of this is inaccurate and misses the point of the question. Ensure you answer the question as set!

Even though there are mistakes and errors, the examiner has attempted to read between the lines. Basic and simple.

A grade candidate – mark scored 15/15

(a) It appears that only a very small area of the photo is low-lying and flat, which clearly makes the establishment of a settlement difficult ✔. The glacial area and ice obvious in the mountains means that the cold might also prevent permanent settlement ✔. It is likely that this whole area experiences tundra conditions. It is hard to develop tundra areas, conditions are harsh and the environment is fragile. Lowland areas may even be marshy in the summer months, because of the seasonal melting of the permafrost ✔. The effect of building on this active layer or thermokarst is to cause buildings to tilt, crack and collapse ✔.

Excellent answer identifies limited lowland, lots of mountains and 'cold'.

Perhaps it misses communications and the lack of any settlement. Concise and uses the photo well.

(b) - During the 1960s and 1970s oil was discovered in Alaska. People want to live here now because they think that they will get rich quick ✔, not just from the oil finds but also from other resources. Others want to work for those companies that are exploring these areas ✔.

Don't spend time thinking about this type of question. It's all facts – you should know them!

- As people have higher disposable incomes, tourism in these areas is increasing ✔. Ecotourism and adventure tourism is on the increase, and people are also needed to provide for these visitors ✔.

(c) There are currently huge conflicts between modern development of Arctic areas and those with a concern for its well being. The development of such areas has far-reaching consequences. Breeding grounds become inaccessible. Hunting of indigenous species increases and road-kill has an effect too ✔✔. Development also upsets the thermokarst landscape, as it alters insolation received by the soil ✔, so raising and lowering permafrost levels. And the environment is in such delicate balance this can be disastrous for animals and vegetation ✔. The traditional lifestyle of indigenous peoples such as the Inuits is disrupted ✔✔. Exposure to Western lifestyles through media exposure and from the visitors and workers ensure that there is a steady stream of young people wanting to leave to further their education and prospects in the USA and Canada proper ✔.

Though not strictly necessary, the first few lines provide the context for the rest of the answer.

An excellent logical consideration of a range of relevant conflicts and their results. Relevant exemplification.

Exam practice questions

A *Answers on p. 33*

Structured question

(1) Study the photograph of a small part of the Torngat Mountains in Labrador, Canada. The cirque shown still possesses a small perennial glacier, though in recent years it has retreated.

Source: Reproduced with permission of the Minister of Public Works and Government Services Canada, 2001 and courtesy of the Geological Survey of Canada.
Photo by: H. Russell

(a) Name the processes at work at A and B on the photograph. How do they differ from one another? **[4]**

(b) Name the landforms at X and Y, and explain how they were formed. **[4]**

(c) The table below shows the orientation of 100 cirque basins in the same Torngat area.

(i) On the diagram, right, plot the cirque orientations.

Orientation	No. of cirques
W	5
SW	4
S	4
SE	20
E	18
N	22
NE	27

[7]

(ii) Describe the pattern of orientation. **[4]**

(iii) What climate factors have affected this pattern of orientation? **[4]**

(d) What do you understand by the terms i. glacial flow and ii. glacial retreat? **[5]**

(e) 'The effects of periglacial processes, though perhaps less dramatic than glacial processes, still produce features that are recognisable in today's British landscape.' Describe some of these features. **[10]**

Synoptic essay

(2) Using examples, discuss the relationship between human activity and the glacial environment. **[25]**

Answers

(1) **(a)** It is likely that freeze thaw is taking place at A (a weathering process). Abrasion (an erosional process) is taking place at B. The rotational movement of material at B forms and maintains the characteristic armchair shape of the cirque.

(b) The feature at X is commonly called the backwall or headwall. It has been steepened by freeze thaw both during and after glaciation. The landform at Y is the rock lip. It forms because of overdeepening of the corrie, by rotational slip. As ice leaves the corrie, by extending flow, there is a lack of glacial energy, the rock lip is left in the landscape.

(c) (i) On the diagram you should have seven correctly plotted lines.

(ii) You should cover the fact that most cirques occur between N and SE, and that there are small numbers of cirques at other orientations.

(iii) It is colder and therefore less melting (ablation) occurs on slopes facing from NE through to SE. These slopes are in shade for the best part of the day. Those in all other areas always face the sun and therefore little snow collects. Wind directions during glacial periods may also have aided snow accumulation.

(d) The weight of ice causes deformation. This deformation causes crystals to slip on planes parallel to the basal plane. This initiates downslope glacial flow. When accumulations of snow fail to keep pace with ablation, the glacial mass responds by retreating up its valley (glacial retreat).

(e) Features include ice wedges, common in the tills of North Norfolk's cliffs. The pingo ramparts (now rounded lakes) of South West Norfolk were former active pingos. The hummocky landscape of North Central Norfolk is the result of thermokarst activity. The patterned ground, stone circles, polygons and stripes are the result of periglacial activity on slopes in Breckland in Norfolk. In higher areas in Britain, screes, blockfields and tor-like features also bear testimony to the effects of periglacial activity. Dry valleys in the southern part of the UK are the final obvious legacy of periglacial activity. The Coldstream area of Scotland has many of these features. Also asymmetric slopes caused by variations in aspect e.g. River Exe, Devon.

(2) *Lowland areas* – soils in glacial and periglacial areas both limit (when soils are thin) and aid agriculture (where thick fertile clay deposits exist). Transport can be aided (following esker and moraine ridges) and be made more difficult (in lake-covered landscapes, e.g. Finland). Other uses include: forestry on glacial sands; sands and gravel deposits are used in the building industry; military trainers utilise the Breckland area of Norfolk because of its low agricultural potential.
Highland areas – Soils are thin, slopes are steep – this, combined with a harsh climate, limits agriculture. Such areas are useful for afforestation, HEP generation, water storage, leisure and tourism.

Questions with model answers

C grade candidate – mark scored 11/15

 For help see Revise A2 Study Guide pp. 37–52

(a)   Describe and account for the main characteristics of the humid, tropical/equatorial climate. **[15]**

The climate at the equator has its own unique characteristics. It is very hot, humid and rainfall is a very regular occurrence ✔. *This is all directly in relation to its position on earth in relation to the sun. The equator is the centre of the earth's surface, i.e. 0°* ✔. *This means that it is always square on to the sun regardless of the tilt of the earth. It is for this reason that the equator and equatorial regions receive maximum insolation from the sun; instead of the sun's heat and light spreading out over the earth when it reaches it due to its curvature it is concentrated in a much smaller area* ✔, *see diagram below:*

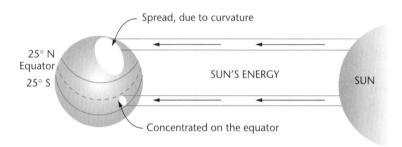

Because this solar energy is much more concentrated at the equator the climate is much hotter than 40°-60° N and S for example where the sun's energy will be more scattered meaning lesser-concentrated insolation and lower temperatures.

As the temperature of the region goes up and the climate gets warmer, moisture towards ground level begins to evaporate. As the moisture evaporates it is taken up into the atmosphere by the very strong convection currents present due to the intense heat. These currents are due to the Hadley Cell which is the name given to the currents between 0° and 30° N and S ✔. *As the water vapour is taken up into the sky it is cooled and condenses once it reaches dewpoint. Cumulo-nimbus clouds form due to the strong upward currents present* ✔. *The droplets grow in size probably through the coalition/collision theory as more water vapour condenses in the cloud. This process happens very quickly and the cloud is ready to rain in a matter of a few hours. When it does rain it comes down in torrents and lasts a few hours. Once the clouds have rained out the sky is clear and the process begins again* ✔. *The humidity of the equatorial climate is due to the regular rainfall but also due to the uplift of moisture caused by the very strong convection currents present.*

Examiner's Commentary

*Part **(a)** attempts to pull together your knowledge and understanding of causal factors. This is effectively done here.*

No!

Avoid sweeping statements that have no credible back-up! Be as accurate as you can.

Diagrams, however simple, save hundreds of words!

✔

Does the examiner need this information?

Back to the point – good use of technical language. Diagram would help! 'Rained out' – too colloquial.

C grade candidate continued

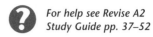

For help see Revise A2 Study Guide pp. 37–52

Examiner's Commentary

The climate therefore becomes very predictable and a typical day in the equatorial climate such as the tropical rainforests in Brazil ✔ (South America) would go like this:

6am the sun would rise and the mist that had formed overnight will begin to burn off. The dew and any other moisture on the ground or vegetation will begin to evaporate ✔. By 8am the temperature will be around the daily maximum and most of the moisture will have gone. The dark, tall cumulo-nimbus clouds will begin to form and it will begin to rain at about 3 to 4 o'clock and will last a few hours ✔. Once this has finished the sky will be clear and the evening will be warm ✔.

Temporal and spatial variations and trends covered effectively here.

A grade candidate – mark scored 9/10

(b)   How true is it to say that equatorial climatic conditions provide an unlimited potential for agricultural growth?   **[10]**

From the vegetation present in the rainforest we can see that the conditions in the equatorial climate are very good for certain types of agriculture. To say they offer unlimited potential is very different. The very high levels of insolation that the equatorial regions ✔✔ received, coupled with the regular rainfall and high humidity means that plants will grow very well as the conditions are ideal for photosynthesis to take place effectively ✔✔. This means that arable farming, potentially, in the equatorial climate could prosper and the crops should be large ✔. To achieve this however is a different story because most of the equatorial regions are TRF (Tropical Rainforest). This means that in order to form the lands, deforestation must occur. The first few years of farming would be good as the soils will be fertile ✔, but because the decaying forest material is no longer replenishing the nutrients ✔ the soils will become infertile meaning that the climate is not enough alone to support the farming ✔.

This is where other parts of the specification are integrated and tested!

Clear understanding of TRF processes.

Only certain animals can survive the conditions, such as intense heat, for example buffalo, which are grazed in the equatorial regions. This has not really changed for many years and is proof that whilst the climate allows the TRF to thrive, it is not really ideal for agriculture and whilst it holds some potential for development it does not hold unlimited potential for the development of agriculture in the equatorial regions ✔.

Makes/draws a sensible and appropriate conclusion.

Exam practice question

Structured question

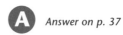 Answer on p. 37

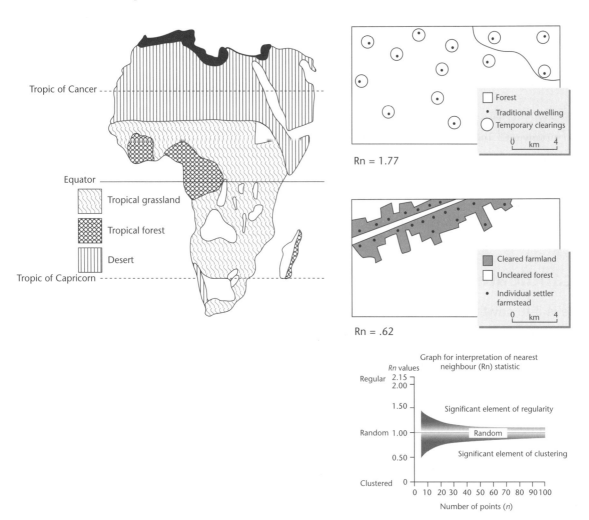

Source for diagram bottom right: Wilson, J, *Statistics in Geography*. Schofield & Sims Ltd., 1995

(1) **(a)** Study the map of Africa above; it shows the distribution of tropical biomes in Africa. Describe the distribution of any of the biomes shown and discuss the climatic nature of their location. **[8]**

Synoptic sub-section

(b) Study the maps of forest-clearance sites above, their nearest neighbour analysis and the significance graph offered.

With reference to the maps and the nearest neighbour statistic (Rn), describe the contrasting settlement patterns and discuss how both traditional and modern clearance systems may limit human impact on this forest system. **[12]**

(c) Using case-study material, compare the tropical forest and the tropical desert ecosystems with reference to their biomass, productivity, tropic structure and nutrient cycling systems. **[15]**

[CCEA]

Answer

(1) (a) Whatever biome is chosen, there needs to be a brief description of the distribution, relating to latitude and other biomes. You could use terms such as tropical, sub-tropical and equatorial. The climate regime is necessary; so comment on seasonality of rainfall, humidity and precipitation.

Tropical forests	Tropical grasslands	Tropical deserts
Distribution – principally 5° either side of equator. **Climate** – high rainfall over the whole year (intense convectional in type). High constant temperature 25°C. Shifting ITCZ has little effect.	**Distribution** – 10° to 15° N, to the north of TRF and south of the desert areas. **Climate** – seasonal. Long, dry, warm winters (25°C) and very hot summers (30°C). Rainfall shifts with the ITCZ. Rain is unreliable and variable.	**Distribution** – across continents, north of the semi-arid grasslands. **Climate** – arid throughout the year. Is beyond the rain belt. Temperature is moderate in winter and very hot in summer.

Examiner's tip

Not a very difficult task. It's all recall and description and only becomes difficult if you haven't prepared yourself properly! The map provides you with a good deal of information – say what you see!

(b) You must apply your good knowledge of your specification to this answer! Plenty of human and physical references are a must.

Map A: Rn = 1.77 Indicates regular spacing of the native settlements. Traditional slash and burn and shifting cultivation exists. Fallow is common. There is low population density. The idea of 'temporary' is important here.

Map B: Rn = .62 This is a clustered distribution. Modern clearance and new routeways mean there is strong linearity. Deep penetration of the forest is unlikely.

Examiner's tip

Many of the specifications require you to have a basic statistical knowledge and to be able to apply it to 'real' situations. You won't be expected to work out calculations, but you should understand the significance of results.

(c) All that is required is a straightforward comparison and something on the nature of the ecosystem and its structure. The A2 book covers both these biomes in some detail – refer to pages 37–50 for help! In the AS book pages 82–87 are useful.

Examiner's tip

In comparison questions you must use words like 'similar'/ 'alike'/ 'different', etc. You should aim to round everything in a sensible integrating conclusion.

Question with model answer

A grade candidate – mark scored 23/25

 For help see Revise A2 Study Guide pp. 54–67

With reference to named desert environments, discuss the role of water, wind and humans in the formation of their landforms. **[25]**

The definition of a hot arid, or desert, environment may be described in various ways but the traditional term is an area with less than 250mm rain/annum. More recent scientific methods of definition are based on the amount of precipitation and water moisture from evapotranspiration ✔. In areas of seasonal drought the term potential evapotranspiration is used where the amount of water available is less than the amount which could be evaporated ✔. The resultant process of aeolian (wind) and water erosion, transportation and deposition have a big effect on the environment and have contributed greatly to the development of landforms in such hot areas, like dunes and desert pavements or reg ✔.

Weathering processes will vary in relation to where the desert is located. Coastal deserts have little variation in temperature unlike internal areas where radiation and lack of cloud cover can make temperatures quickly fall ✔. Despite the absence of water, inland deserts can have higher weathering rates than the corresponding coastal ones. Freeze thaw can be common here due to poor vegetation cover ✔. Insolation weathering was thought to cause the surface layers of the rock to peel off by exfoliation or individual grains to break away by granular disintegration ✔. As a result of this, cracks can appear in the rocks by pressure release as the rock loses mass. In areas of rock with chemical weaknesses such as granite, chemical weathering may still occur despite the very low precipitation levels. In deserts there is often sufficient moisture to cause certain rocks to be affected by hydration or to peel by exfoliation (a combination of chemical and physical weathering).

A second cause of weathering is salt weathering. Salts in rainwater form crystals because moisture is readily evaporated in the high temperatures and low humidities ✔. The salt expands and mechanically breaks off pieces of rock in which they have formed. Salts that accumulate near the desert surface are thought to form duricrusts. Desert varnish is a dark glazed surface that has been coated by iron or magnesium oxide following the evaporation of capillary water.

Aeolian processes can be considered the most important factor due to limited vegetation to bind surface layers together. Two processes make up aeolian processes: deflation and abrasion. Deflation is the removal of fine material by the wind leaving a pebble-strewn desert pavement or reg ✔. Abrasion is the sandblasting action effected by materials as they are moved by saltation. This

A grade candidate continued

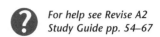

For help see Revise A2 Study Guide pp. 54–67

Examiner's Commentary

process smoothes, pits, polishes and wears away rock close to the ground. Abrasion produces a number of distinctive landforms including rock pedestals, ventifacts, yardangs and zeugens ✔.

Good use of vocabulary.

Aeolian transport processes are greatest where winds are strong, turbulent, come from a constant direction and blow steadily for a long period of time. The optimum conditions for wind transportation are found in these arid and semi-arid environments. The three main processes at work are suspension, saltation and surface creep ✔.

Landforms are greatly influenced by deposition by the wind. It is mainly lightweight particles, which produce dunes. Some dunes are formed around obstacles others form on quite even surfaces. A variety of different dunes have been identified, principally from air and satellite photography.

The other factor that has a big impact on arid environment landforms is water ✔. Rivers in arid areas fall into three main categories: exogenous rivers such as the Nile, which flow through the whole year, even if discharge is reduced by evaporation. Endoreic drainage, where rivers terminate in inland lakes, such as the River Jordan flowing into the Dead Sea and ephemeral streams which are seasonal but can develop high levels of discharge due to torrential seasonal rain, inhibited infiltration and desertification ✔. Sheet floods can occur where water flows evenly over the land and is not confined to channels. Much material covering the desert floor is believed to be deposited by this method during earlier wetter periods called pluvials ✔. Very soon the collective run-off becomes concentrated into deep, steep-sided ravines known as wadis which are prone to infrequent flash floods.

During pluvial periods the climate was wetter than at present.

Today there is evidence to suggest that people grew crops in many arid and semi-arid areas.

Stretching from the foot of the highlands is often a gently sloping area of base rock or covered in a thin veil of debris - a pediment ✔. There are two theories of origin: one claims that weathered material from the cliff faces or debris from alluvial fans was carried during pluvials by sheet floods. The sediment passed over the lowlands before being deposited, leaving a gentle slope of some 7°. The alternative theory involves the parallel retreat of slopes resulting from weathering ✔. Playas are often found at the lowest point in the pediment. They are shallow, ephemeral, saline lakes formed after rainstorms. Heavy evaporation leaves flat layers of sedimentation. If clay, large dissected cracks up to 5cm deep form ✔. Occasionally, isolated, flat-topped remnants of former highlands rise sheer from the pediment. Mesas in Arizona have summits large enough for the Hopi Indians to have built their villages thereon ✔. Buttes are similar smaller versions of mesas.

There is a contribution from wind to pediment formation.

Questions with model answers

A grade candidate continued

For help see Revise A2 Study Guide pp. 54–67

Examiner's Commentary

Unfortunately, man, has attempted to capitalise in the available resources of arid environments. Dams have been built that reduce the risk of loss of human and animal life as well as maintaining lakes sufficient to rear fish and for transport ✔. They enable crops to be grown in areas or at times of year where the climate was too arid. However, the building of dams has often had an adverse effect because by altering discharge, sedimentation levels are affected e.g. the Nile. Elsewhere the reduced discharge has encouraged deposition, braiding and the raising of riverbeds. Sediment-free water has also, in some desert areas, encouraged the rapid growth of floating plants, such as water hyacinth, which has blocked channels ✔. The creation of lakes behind dams can alter local climates, e.g. reduction in the discharge of the River Ob in Russia might accelerate the formation of ice so that it freezes up to 19 days earlier than is normal. There are of course other changes associated with changing land use.

> *In a synoptic question you should expect to go beyond mere description – this essay certainly does this!*

Many of the recent changes have been shown by satellite photos, especially in the African deserts where there has been an increase in dust storms, the number of dunes and the areas of wind-blown bedrock or yardangs ✔. Areas where wind appears to be the dominant geomorphological agent are known as aeolian domains, although it is still debatable whether such features as dunes are active or form fossil features ergs. Fluvial domains are where water processes are dominant or, as evidence suggests, were dominant in the past ✔. Landforms produced by each co-exist in the same area, but the balance between their relative importance has changed significantly over time ✔. At present it would appear that the role of water is declining while that of the wind is increasing, compounded by human mismanagement of areas where the environment equilibrium is very fragile ✔.

> *If there is a fault in this essay, it's that actual desert areas aren't mentioned enough.*

Exam practice questions

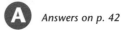
Answers on p. 42

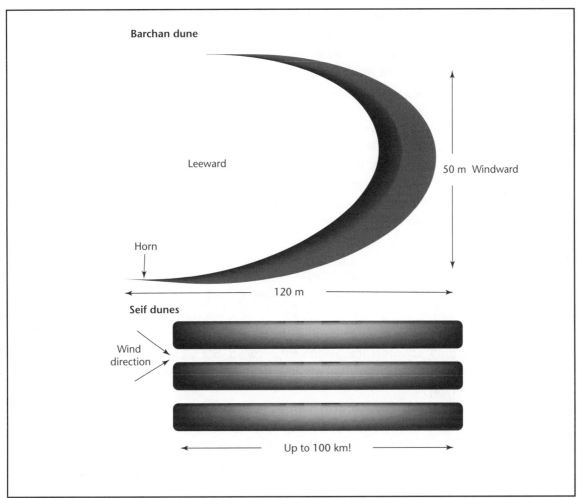

Structured question

(1) **(a)** Explain the plan section of either the seif or the barchan dune. **[3]**

(b) Draw a profile of the barchan dune and add labels to explain how it formed. **[6]**

(c) Explain why barchan dunes are seen as a threat to human settlement. **[4]**

(d) What has man done to reduce the threat that such dunes pose? **[4]**

Synoptic essay

(2) Describe and explain how the impact of irrigation processes influences population, settlement and economic development in desert environments. **[25]**

Answers

(1) **(a)** **Barchan** – Pebbles or an obstruction might retard saltated sand. Wings/horns are swept forward. Further sand arrives and is shepherded onto the flanks and runs off in streams. Turbulence on the leeward side sweeps the barchan court clear. Sand replaces that which is lost. Winds must blow almost constantly in one direction for barchans to form.

Seif – Has a regular cross-sectional form and orientation. Parallelism is common. There is a correlation between corridor width and dune width, and between dune height and spacing. Shape is the result of complex annual wind regimes. 'Speculation has it that large-scale rotary movement in the air stream' produces the regularly spaced thin sand strips; they are maintained by 'transverse instability', i.e. strong wind is decelerated at ground level and sand is transferred to the 'strips' – Bagnold. Others suggest that seifs may be the result of erosional activity.

Examiner's tip

Expect this sort of question. It tests not only your vocabulary but also your process understanding. Candidates frequently misread questions; this, for example, is an either/or question!

(b) Labels to include: unanchored, asymmetrical, windward, leeward, horn, crest, steep/shallow slope, direction of wind, direction of movement, apron and streamer. Explanation as above in **(1)(a)**.

(c) Increasing use of marginal lands means that man has had to inhabit the desert areas of the world. Barchan dunes (transverse dunes) are the most mobile of dunes; they are said to have a 'short memory'. Sand drift in barchan dune areas is rapid (e.g. Lüderitz, Namibia). Barchans form, grow and migrate, in some cases in hours. Peruvian barchans move on average 1.7 km in their 64-year lifecycle. Barchans don't travel alone!

Examiner's tip

Applying your accumulated knowledge is what it's all about. Here man's influence is again examined in the arid zone. It's important that examples flower your answer here! The examiner will look for four points or two well-developed ideas.

(d) Dune stabilisation – with vegetation. Planting 'sand breaks' – of eucalyptus for instance.

(2) Typically this synoptic question links man into a geomorphological process. It asks you to show how irrigation in the desert realm has allowed populations to grow, and settlement and economic development to take place. You might reference the process of population growth to development. In linking the above parts together, you will make the connections to other parts of your specification. You have, of course, to demonstrate an understanding of modern (field channels, canals, pipelines, drip irrigation, boom irrigation, pump irrigation) and traditional (annual flood, the shaduf/Archimedes' screw) irrigation processes. Some reference to actual examples will, of course, be important e.g. Alice Springs in Australia's Northern Territory.

Examiner's tip

Your ideas must flow in these synoptic essays. It's a reasonably straightforward question, the content of which is sometimes overlooked when studying deserts. In many ways this is where the primary 'link' exists, i.e. man's interaction with the desert.

Question with model answers

C grade candidate – mark scored 6/10

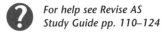

For help see Revise AS Study Guide pp. 110–124

Extended writing

Explain how physical factors can lead to large areas of low density of population. **[10]**

[AQA specimen, 2001]

Examiner's Commentary

People do not want to live in areas of very low annual rainfall like the Sahara Desert. In these areas there is going to be a lack of water to support the population. Crops would be difficult to grow because of insufficient water, poor soils, and the extremes of temperature that dry crops during the day and freeze them at night ✔.

A general introduction would be good here; one that refers specifically to the question.

Large area like the Amazon Basin where there is a dense rainforest cover, and the weather is hot and humid will also not attract a large population ✔. The Amazonian soils are also thin and easily leached when cleared for farming, due to the heavy tropical rainfall ✔. Large areas of the basin are also experiencing severe soil erosion because of deforestation and overgrazing by cattle ranchers.

Good understanding of processes.

Areas devoid of natural resources and mineral wealth are unlikely to attract large populations, because many raw material extractions can support not only the miners, but also the support services and could provide a good energy source in the form of coal ✔. Towns such as those in the Rhondda valley can develop in such areas.

People want to live in areas of natural beauty, where all physical features are balanced; the climate is predictable and suitable for living. Florida suffers from frequent hurricanes and this is the reason why it is historically sparsely populated ✔, even though people today tend to ignore such hazards for the natural beauty of the area.

This shows a good understanding of how some modern settlements have developed despite the poor physical conditions.

Mountainous areas of the country are often sparsely populated because construction and maintenance of not only homes but also transport links are more expensive and harder to maintain. Areas where communications are good, and have the potential for natural ports attract people because these don't only provide jobs but also their quality of life is likely to be better ✔. Trade is better; exports and imports mean that goods are more readily available and diets may be better and more varied.

This piece needs a more thorough conclusion, and could be improved by using simple case studies to highlight the points.

Therefore physical factors can lead to large areas having a low population density.

Question with model answers

A grade candidate – mark scored 9/10

 For help see Revise AS Study Guide pp. 110–124

Physical factors tend to have scope for leaving large areas without population and have a much greater impact than human factors do at such a scale ✔.

There are many physical factors that influence distribution, but on closer observation of a map of world population distribution, the dominant factor appears to be temperature. In areas of excessively high temperatures such as the Sahara, crop growth is difficult. The high rates of evaporation mean that water barely has time to influence the development of soils ✔. Areas such as the Amazon Rainforest also have poorly developed soils, which are easily leached when cleared and have little capacity to hold water ✔. The implications for human populations are huge as water and food supplies are fundamental to survival and permanent settlement. There are exceptions to this rule, as there are many areas of the Middle East that, although desert, are highly populated due to the presence of rich reserves of oil ✔. Areas of extreme cold can also be sparsely populated. There are vast tracts of land in Northern Canada and Alaska where the cold climate cannot support any more than a few people ✔. Here again, oil resources have meant that there are some areas of dense population density.

Other sparsely populated areas are ones where the topography of the land is difficult and the gradient steep. This is clear in the interior of Japan where the mountainous terrain and thin soils are sparsely populated and farmed ✔, as opposed to the coastal plain where there are high density conurbations and very intensive agriculture. Large mountain ranges such as the Himalayas may also have extremes of temperature and many areas where the altitude prohibits human settlement ✔. The settlement that has taken place in many areas of Nepal has caused deforestation and intensive soil erosion, and thus prohibited the development of high population densities ✔.

The above factors are perhaps responsible for the greatest wastelands on earth, but there are factors that discourage population growth. Many parts of the Tropics are malarial and this kills over a million people a year ✔. There are many factors that discourage high densities of population, but with the growing world population, such problems will have to be overcome if we are to continue to live upon the earth.

Examiner's Commentary

Clear introduction, explaining the reason for the question being asked in such a way.

Good linkage of ideas; temperature and then soils.

Excellent highlighting of exceptions and therefore balance are vital to geography writing.

It would be good to discuss aspects of natural vegetation here.

Good use of technical language.

Other hazards could be considered here.

An excellent summary conclusion which suggests that the student understands the bigger picture.

Exam practice questions

Structured questions

Answers on p. 47

(1) **(a)** Complete and label the Demographic Transition Model which has been started in Figure 1. **[3]**

(b) Figure 2 shows the average population values for the five continents in 1985 and 1995.

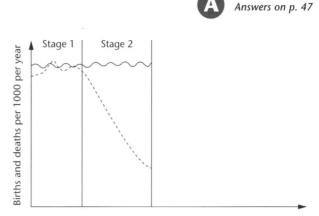

Figure 1: The Demographic Transition Model

	1985			1995		
Continent	B.R.	D.R.	Fertility Rate	B.R.	D.R.	Fertility Rate
Africa	45	16	6.3	42	18	5.9
Asia	28	10	3.7	25	8	3.1
Europe	18	10	1.8	12	11	1.6
N. America	15	8	1.8	16	9	2.0
S. America	31	8	4.2	27	7	3.2

B.R. = birth rate per 1000 people per year
D.R. = death rate per 1000 people per year
Fertility Rate = average number of children born to a woman during her life time

Figure 2: Population statistics by continent

Using the data in Figure 2, state the Demographic Transition Model stage in which you would place

(i) Africa

(ii) Asia **[5]**

(c) Explain why the greatest declines in fertility rates between 1985 and 1995 were in the continents of the less economically developed world. **[7]**

[AQA]

(2) Figures 3 and 4 show population pyramids of the total population of Greenland and the emigrating Greenlanders. Greenland is a part of the Danish State, and therefore has close economic, political and social ties with it. Its total population is 56,083 (1999) at a density of 0.14 km². Natural increase is at 1.4% and net migration is –0.9%.

Figure 3: Total population of Greenland 1996   Figure 4: Emigrating Greenlanders 1996

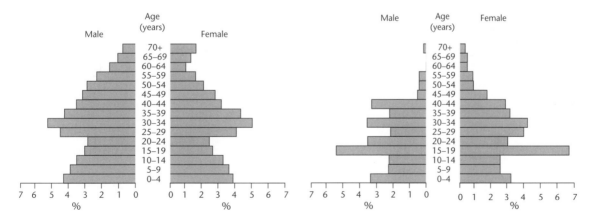

(a) (i) Name one age group that has been affected by out-migration. **[1]**

Exam practice questions

    (ii) State and explain the evidence from Figure 3 and Figure 4 that suggests that this age group has been affected by out-migration. **[6]**

    (iii) Suggest one possible reason for the out-migration of this age group. **[4]**

  (b) Describe one potential impact of this out-migration on:

    (i) the origin (Greenland)

    (ii) the destination **[8]**

  (c) Using evidence from Figures 3 and 4, suggest how and why the age structure might change in Greenland in the next 30 years. **[11]**

(3) Figure 5 shows the projected population growth for the LEDCs and MEDCs of the world between 1950 and 2150.

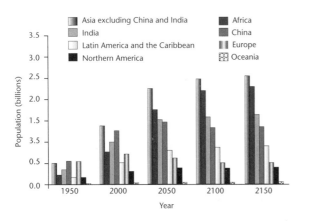

Figure 5: Long-range world population projections, based on the 1998 revision

Source: Adapted from Population Division of the Department of Economic and Social Affairs of the United Nations Secretariat (2000). *Long-range World Population Projections: Based on the 1998 Revision* (United Nations publication, Sales No. E. 00. XIII.8).

  (a) State and explain the differences between the projections for population growth between the areas. **[8]**

  (b) Define the following terms:

    (i) overpopulation

    (ii) underpopulation **[4]**

  (c) What are the likely consequences of overpopulation on LEDCs? **[7]**

  (d) Outline the possible solutions open to individual nations to deal with overpopulation. **[11]**

Answers

(1) (a) Complete the DTM Stages 3 and 4 using the same symbols labelling the lines or adding a key. Death rate is always the first to fall in Stage 2, followed by the birth rate in Stage 3, both being low and fluctuating in Stage 4.

(b) (i) Africa would be in Stage 2 as there is a large gap of 29‰ between BR and DR (**natural increase**), which remains high after DR has fallen.

(ii) Asia is more likely to be in Stage 3 as the DR is low at 10‰ and the BR is declining at 28‰, which is clearly declining but not low when compared to Europe at 13‰. With a slightly higher BR and DR it might be in Stage 2, and with a slightly lower BR it could be in Stage 4, but is not as the natural increase is still high.

(c) These areas have undergone any of the following recent developments: technology, education, medicines or social changes. Develop this by adding that these changes are often forced upon governments by the pressure on resources that population growth can cause.

(2) (a) Try to pick a **cohort** (age group) from the population pyramids. If you chose the 15–19 category you would need to use precise figures from both the graph, which shows the indentation in the overall population structure, suggesting that this age group might have emigrated, and the data from the emigrating population structure, which shows the high numbers of emigrants from this cohort. Between 15 and 19 years old, they are likely to leave to pursue further education or employment opportunities due to Greenland having only small urban settlements and therefore few opportunities when compared to those on offer in Denmark.

(b) (i) Most likely impacts are social and economic.

(ii) It can only have a negative impact on the Greenlander culture, and cannot help businesses that might need bright young workers, and vice versa for the destination, with increasing competition for jobs and dynamism in the economy.

(c) Use specific percentages, there are many impacts, a greying population, rural depopulation, social welfare and pension provision.

(3) (a) Describe to enable you to explain
  • slow growth or stagnation in **MEDCs** with low/declining birth rates and low death rates
  • high sustained growth in **LEDCs** with declining death rates and huge proportions of the population of child-bearing age.

(b) (i)/(ii) Too many/few people for available/potential resources with given technology.

Examiner's tip

*It is important to understand both the ideas of **Malthus** and **Boserup** in relation to population and resources as well as to one another. Their ideas are tied to the time they wrote them and have both applications and limitations.*

(c) **Malthus** or **Boserup** are good to mention here, pressure on resources and environment, conflict over resources, migration, famine, technological change etc.

(d) Can be argued two ways but you should use at least two examples: population as problem, therefore anti-natal policies and population as solution/resource, therefore develop economically.

Examiner's tip

*When studying a migration stream it is important to understand their motives and obstacles to their migration, but also the impact that their move may have had on the **origin** and **destination** locations. These could be environmental, social or economic consequences, and might help you to understand any **return migration** that might take place.*

Question with model answers

C grade candidate – mark scored 6/10

 For help see Revise AS Study Guide pp. 90–106

Examiner's Commentary

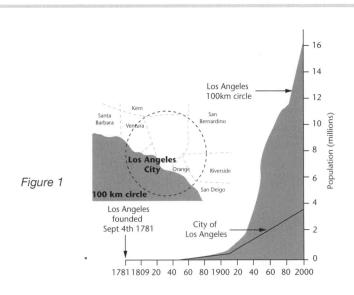

Figure 1

Los Angeles 100km circle

Kern

Santa Barbara

Ventura

San Bernardino

Los Angeles City   Orange   Riverside

100 km circle   San Deigo

Los Angeles founded Sept 4th 1781

City of Los Angeles

Population (millions)

1781 1809 20 40 60 80 1900 20 40 60 80 2000

Study Figure 1, which shows the growth of Los Angeles in the USA.

(a)   Describe the pattern of growth of Los Angeles.   **[3]**

(b)   Use Figure 1 and your own knowledge to explain how Los Angeles can be described as a mega-city.   **[2]**

(c)   Many people argue that mega-cities such as Los Angeles are unsustainable. Explain why.   **[5]**

[Edexcel specimen]

(a) Since its founding in 1781 LA has enjoyed steady growth from 1850 to 1910, reaching a population of about 1 million. After about 1910 the population rose quickly and has reached a population of 16.5 million today ✔, slowing only briefly in the 1970s.

(b) LA can be described as a mega-city because, as we can see from the graph and the map, it has a large population and also because it covers a large area of land ✔.

This C grade answer needs to be precise about the definition of a mega-city and also how big LA is.

(c) LA can be described as being unsustainable because it is a huge city that relies heavily on the car ✔ and thus a great deal of energy is consumed and pollution produced by each person driving all over the city to each of the different activities that they might do in an average day ✔. Many mega-cities such as London are also great consumers of manufactured goods and processed food ✔, and due to this produce a great deal of waste that has to be dealt with. They also require huge quantities of energy that is difficult to obtain from renewable resources, so a great deal of pollution is inevitable ✔.

The C grade answer has good coverage of points but needs to use more specific case-study information.

A grade candidate – mark scored 9/10

For help see Revise AS Study Guide pp. 90–106

Examiner's Commentary

(a) Los Angeles was founded in 1781, but it was not until 1900 that its population has grown to $\frac{1}{2}$ million people ✔. This slow growth changed after 1910 when the population rapidly rose from $\frac{3}{4}$ million to 2 million in only 20 years ✔. The rate of growth changed again after 1930 when an exponential growth curve saw the population increase 4 fold by the mid-1950s ✔. After this the growth rate reduced, especially during the late 1970s, but today continues to grow and stands at 16 $\frac{1}{2}$ million people.

Here the A grade answer is specific about both dates and figures.

(b) Los Angeles can be described as a mega-city primarily because it has a population of 16.5 million, well over the 8 million threshold that would make it a mega-city. It also takes up a huge area. Cities as large and complex as LA tend to have multiple nuclei and little internal cohesion ✔.

Excellent use of the definition, but also great extension with the multiple nuclei idea.

(c) Mega-cities can be considered unsustainable as many of them do not practise the principles of sustainability. A sustainable city should try to reduce its waste and recycle what it produces ✔. It should also try to use energy more efficiently by making public transport efficient and reducing dependency on the car ✔. LA is the ultimate motor city with major roads linking all amenities ✔. It has been said that Californians would travel to the toilet in their car if only it had valet parking ✔. However it could be argued that the real differences are between cities in MEDC and LEDC. Mega-cities such as Bombay do recycle their waste and utilise their inputs very efficiently ✔. Ultimately what mega-city can continue to function as it does without harming the environment for future generations?

The A grade answer presents a balanced argument, very important in geography.

An excellent conclusion that takes us back to the question.

Exam practice questions

Structured questions

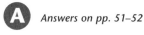

Answers on pp. 51–52

(1) (a) Study Figure 2, which shows the average annual % rate of urbanisation 1995–2000.

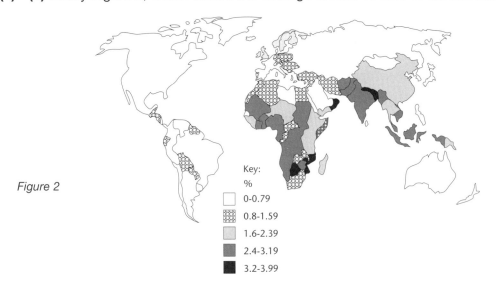

Figure 2

Key:
%
☐ 0-0.79
▨ 0.8-1.59
▦ 1.6-2.39
▨ 2.4-3.19
■ 3.2-3.99

Source: Clark, C, Matthews, H & Binns, T, *Geography Review A-level resource pack –
Urban Development and Change*. Philip Allan Publishers, 1998

(i) Annotate the world map above to show the differences between rates of
urbanisation between continents. **[6]**

(ii) Select two continents and outline the reasons for the differences in rates of
urbanisation. **[6]**

(b) For what reasons does counterurbanisation usually exceed urbanisation in more
economically developed countries? **[8]**

(c) Outline implications of:
either counterurbanisation on rural settlements in MEDCs
or reurbanisation on inner-city areas in MEDCs. **[10]**

(2) (a) Describe the land use of any named LEDC city. **[7]**

(b) Explain the limitations of the application of any model of urban morphology to
the city you have described in **(a)**. **[7]**

(c) With reference to poorer residential areas of LEDC cities:

(i) Outline the environmental, social and economic problems that can be
experienced in poorer residential areas. **[9]**

(ii) Suggest ways in which:
either the local population can help themselves to overcome such problems
or city decision-makers can solve their problems for them. **[7]**

Extended question

Figure 3: Norwich CBD, an urban environment

(3) (a) Using Figure 3, describe and explain
the differences between urban and rural
environments.

[10]

(b) For what reasons are many rural
settlements in MEDCs exhibiting
the characteristics of urban
settlements? **[10]**

Source: The Photographers Library

Answers

(1) (a) (i) Concentrate your answers on the differences between the continents, as they clearly exhibit very different characteristics. Use figures and countries to show your geographical knowledge as well as your interpretation skills.

(ii) Pick two **contrasting** continents like Africa and Europe or South America. Clearly Africa and Sub-Saharan Africa in particular are the areas experiencing high rates of urbanisation and this might be due to the recent growth in importance of the major urban centres, and the colonial legacy that means that many capital cities are also gateway cities for exports. In contrast Latin America has apparently smaller rates but this might be due to the fact that urbanisation here began earlier in the 1950s with rapid industrialisation, and most of the nations have very high urban populations already.

(b) Focus on Europe or North America where urbanisation was much earlier and as most nations are now post-industrial economic, having lost much of their heavy industry. Transport and communication developments have enabled people to live further afield; thus people are choosing to live in rural areas rather than urban ones. Use examples of places you know here, stressing both the factors which are pushing people out of the city, and the factors pulling people to the countryside.

(c) Include at least one case study here. For **counterurbanisation** in rural areas you could consider social implications such as loss of community spirit, economic ideas such as increasing house prices due to competition, or environmental issues such as housing developments on greenfield sites. Reurbanisation too can be looked at in a similar way and you should make sure that you understand why people return to the city and what role redevelopment and gentrification play in changing inner-city environments. Also look at positive and negative implications of the process.

(2) (a) The key here is to learn a simplified land-use map of your settlement morphology case studies with names and some statistical data. If you use a well-known example such as Mexico City you need to be much more precise about the detail, than if you use an uncommon case study such as Johannesburg.

(b) Remember that the easiest models to criticise are the most generalised, e.g. Hoyt and Burgess, and if you use an area specific model, e.g. Griffith and Ford for Latin American cities, you might be able to present a more balanced and in-depth argument.

Answers

(c) (i) The structure has been given to you, as has the instruction 'with reference to', so you should try to use as much place-specific detail as you can, the more in-depth the better. Environmental problems might be due to the building of poorer housing areas on **marginal land** such as steep hillsides or marshy areas, with the inevitable landslides and diseases. Social problems are usually to do with a lack of **amenities** such as electricity and sewerage, but could also be due to a lack of **facilities**, leading to poor health and sanitation. Economic problems are usually connected with jobs and the lack of **formal employment** opportunities.

Examiner's tip

Try to read the question in full first, then you will be able to see the links between questions and use your answer in one part to help you with the next.

(ii) Problems and solutions are usually linked. Tie in your case studies from the previous section and consider the structure you have been given. The local population can help themselves through a range of self-help schemes, small-scale industrial developments and, most powerful of all, **collectivisation**. Groups get together to provide amenities, build roads or affordable houses or even to acquire land. From a decision-maker's point of view, they have to achieve the same aims of improving the environmental, social and economic situation, but they have greater powers. Usually cooperation between groups is essential, but decision-makers can do many things to make housing better by giving over ownership of the land, improving facilities and amenities in the planning stages of development and providing transport links to get to work.

(3) (a) Use the picture as a starting point here, but you should certainly brainstorm your ideas as rural and urban are very difficult to define precisely without using a range of different criteria. You should consider ideas to do with density and height of buildings, functions of buildings, land prices, infrastructure and public transport, employment, age structure. Some more alternative ideas such as parking restrictions and street lighting are clear on the picture so should be used in contrast with the other ideas. You then have to explain them, and it is important to touch on the ideas behind bid rent, industrial location and migration, and perhaps even make some reference to historical reasons.

(b) Again, an understanding of ideas such as the model of a suburbanised village, or a case study of a village you know that has experienced such change, might give you a framework to answer this question. Most villages are changing due to an influx of new residents and this leads to a change in the social make-up, increased demand for housing and different demands for services and functions.

Examiner's tip

Remember to answer the question set rather than the one you would like to have been set. It would be easy to answer this question by talking about the reasons for counterurbanisation, but you need to focus on the changing nature of villages.

Question with model answers

C grade candidate – mard scored 9/15

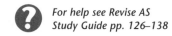

For help see Revise AS
Study Guide pp. 126–138

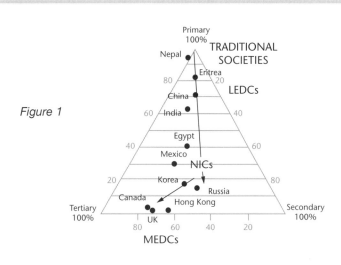

Figure 1

Study Figure 1, which shows a triangular graph of employment structures (1999) for 11 countries at different levels of economic development.

(a) Account for the terminology used to classify the countries in Figure 1. **[6]**

(b) With reference to one region within an MEDC, comment on the view that the changing employment structure has had a positive environmental and socio-economic impact. **[9]**

(a) The terminology used on the triangular graph describes the different types of employment structure that exist for different countries around the world. It begins with traditional societies which are those which have a high percentage of primary industry such as farming or forestry, and their economy is based around the production of primary goods for subsistence or export ✔. Such countries are called LEDCs or Less Economically Developed Countries. The more secondary industry they have the more industrialised they are, growing from RICs or Rapidly Industrialised Countries ✔ to NICs or Newly Industrialising Countries, to MEDCs. MEDCs have less manufacturing industry and more tertiary ✔.

(b) South Wales is an area that has deindustrialised due to the loss of the coal mining industry and the rationalisation of the steel industry under the Conservative government in the 1980s ✔. This left the environment of the valleys scarred and damaged. It is only today after many years of landscaping that many former coal mines and steelworks are not an eyesore ✔. An example of one such area is the former Garden Festival site on the site of the former steel works at Merthyr Tydfil ✔. The economy of the area is still in a very poor way, with rates of unemployment up to 35% and social deprivation levels some of the highest in Europe ✔. Many of the shops are

Examiner's Commentary

An excellent description of the terms, but this candidate needs to use figures, terminology and examples.

Try to avoid political opinion; a geographer should be objective.

The candidate should include detail about the whole region and not focus on one small area.

Questions with model answers

C grade candidate continued

 For help see Revise AS Study Guide pp. 126–138

Examiner's Commentary

closed down and derelict through loss of trade as many people no longer have a high disposable income to spend upon luxury goods and services ✔. This has not encouraged the young people to stay so many have left for the cities on the coast like Cardiff and Swansea, and the area has stagnated due to a lack of dynamism ✔. The environment may have been improved but the life has been taken out of the valleys, leaving them desolate and economically sterile. The overall social and economic impact has been a very negative one.

Try to use geographical language rather than emotive language.

A grade candidate – mark scored 14/15

(a) The terminology used in Figure 1 charts the developmental stages of employment structures from the traditional economies of Nepal and Eritrea to the post-industrial economies of the UK and Canada ✔. Nepal is said to be traditional due the high proportion of people employed in primary industry at 94% ✔. As country becomes more developed it industrialises ✔, and thus the amount of manufacturing industry grows from an RIC (Rapidly Industrialising Country) like India at 16% ✔ to an NIC (Newly Industrialising Country) like Russia at 57% ✔. The post-industrial society sees a greater concentration upon the tertiary sector which is usually associated with deindustrialisation, so that Canada now has only 23% secondary industry but 75% tertiary ✔.

Good clear introduction.

Excellent use of support from the graph.

Good terminology.

(b) The employment structure of South Wales has changed radically over the past 30 years with a global shift ✔ in manufacturing and cheap imports causing the loss of the coal mines and the bulk of the metal works, and the jobs that they created ✔. This deindustrialisation created an initial demultiplier effect ✔ upon the region leading to rates of unemployment reaching 35% in some areas such as the Rhondda valley ✔. This has not had a positive economic impact overall as there is now very little money to go around and as a consequence many of the services and support industries have also left ✔, causing some out-migration to London and the South East of England ✔.

Good use of simple statistic to support the answer.

The Welsh Development Agency then set about trying to attract industry to the area and was given funds from central government to do this. Many industrially blighted areas have been redeveloped and landscaped such as the Garden Festival site at Merthyr Tydfil ✔, and many TNCs such as Ford and LG have been attracted back to South Wales ✔. This has been in no small part due to the extension of the M4 and has seen development limited to the areas adjacent to the motorway like Sony at Bridgend. Thus the changing employment structure of South Wales has served to improve the environment in many areas, but has served to increase spatial inequalities of wealth and service provision.

Good understanding of the processes, but what powers did the WDA have?

Clear conclusion referring back to the original question.

Exam practice questions

Structured questions

A *Answers on pp. 56–57*
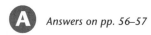

(1) (a) (i) With the aid of examples, describe what is meant by 'a global product'.

(ii) Explain the rapid growth in the number of such products in recent years.   **[8]**

(b) Describe 2 ways of assessing the impact of transnational corporations (TNCs) on the economy of their host country.   **[5]**

(c) Many TNCs have their headquarters in MEDCs but have built overseas branches in NICs. With reference to one or more examples, explain why these corporations have been attracted to NICs.   **[7]**

(d) Discuss the economic and social effects that could occur in the country of origin of TNCs when they make decisions to locate branches in other countries.   **[5]**

[AQA]

(2) Study Figure 2, which shows the relative percentages of world car production in 1970 and 1997.

(a) Summarise the differences between the relative car production figures for NICs for 1970 and 1997.   **[4]**

(b) Explain the significance of the majority of MEDCs having a relatively lower share of world car production in 1997 when compared to the 1970 figures.   **[6]**

(c) Evaluate the extent to which government policies have changed the location of an industry in recent years, within a country of your choice.   **[10]**

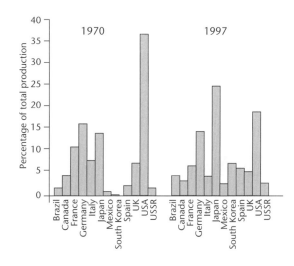

Figure 2: Source: *Geofile 367* (Stanley Thornes, January 2000), © Nelson Thornes

Extended question

(3)

Figure 3

(a) Outline the advantages of globalisation to a large Trans National Corporation (TNC).   **[10]**

(b) Figure 3 shows the views of one anti-capitalist organisation in relation to the activities of a large TNC operating at a global scale. With reference to at least one NIC, evaluate the socio-economic impacts of TNCs operating in that country.   **[15]**

Answers

Examiner's tip

At AS level you should try to integrate case studies and examples into your text, rather than just tagging them onto the end of answers to short questions.

(1) (a) (i) A global product is one that is marketed globally or contains components from a variety of countries. You should include an example here, e.g. Nike trainers or Sony televisions, as well as in the next question.

(ii) Remember that TNCs have huge power to produce, transport and market their goods anywhere. They are able to choose the least-cost locations to manufacture components, then take advantage of better communications to transport them to their country of sale for assembly. This may have the advantage of bypassing many trade tariffs like those in place around the EU. Then with global marketing (using media and the Internet) such products can be sold anywhere in the world, and increasingly so in LEDCs where purchasing power is increasing.

(b) You only need to describe here, and you could choose from ideas such as:
• % of workforce employed by TNCs
• contribution of TNCs to GNP/GDP
• change in export base from primary goods to manufactured products.

(c) You need to base your answer around your examples, also refer to either TNC choices or a specific country's attractions. There are a whole range of ideas here – cheap labour, resources, expanding markets, strong government support, established trade links, relaxed environmental laws – but they must be developed beyond the statement and example to get the better marks.

Examiner's tip

*Two important concepts in industrial change in MEDCs are **deindustrialisation** and **tertiarisation**. You need to understand what the processes are as well as the economic/social/environmental impacts on a national and local scale.*

(d) This is a classic question and could equally be focused on the host, as well as including environmental effects too. The key to this question is to detail the decline of manufacturing industries and loss of associated jobs, e.g. car manufacturing in the West Midlands, UK, and the change to tertiary, quaternary and quinary employment. You should therefore include details of where the new industries move to, e.g. Cambridge, and how the government might encourage growth in areas of decline, e.g. Nissan Factory in Sunderland.

Examiner's tip

Check the scale on graphs carefully, for example in Figure 2 the vertical scale is the percentage of total production. Therefore you can only comment on the relative production figures and not on the actual figures, i.e. South Korea may be producing the same number of cars, but if global production has dropped, then its percentage share will have increased.

(2) (a) Firstly identify the NICs, which are considered to be Brazil, Mexico and South Korea, then use accurate measured figures to summarise their increasing share of global production, e.g. Brazil has increased its percentage production from 1% in 1970 to 4.5% in 1997.

(b) Identify the pattern, then use the concepts of **globalisation** of the car manufacturing process and the subsequent loss of much of the car manufacturing base.

(c) Choose your case study carefully here, but examples could range from iron and steel or car manufacturing to high technology industries or tourism. Make sure to consider the following ideas: planning and policy, taxes and incentives, grants and loans, nationalisation. For example, the British car industry was established principally in the Midlands and London; this changed after the government sold off British Leyland and then offered incentives for foreign car manufacturers to establish car plants in the NE and south of the UK.

Examiner's tip

*It is very important to understand the concept of **globalisation**. It is a strategy employed by TNCs that enables them to take advantage of worldwide opportunities based upon the competitive advantage offered to them by the country they site in, e.g. Toyota built a branch plant in Derbyshire, UK, to enable them to get around high EU import tariffs as well as to take advantage of the incentives being offered to them by the British government.*

(3) (a) Better to use a variety of TNCs here that can provide you with a range of competitive advantages. You could consider the global advantages of:
• economies of scale
• flexible global production strategies
• global marketing strategies
or advantages related to a choice of specific questions:
• environmental issues: lax laws
• labour and unions: cheap, skilled, educated, compliant unions
• markets and trade: close to large market, within trading zone
• sites: cheap land, expansion sites, good communications
• state: low tax, tax-breaks, tax holidays, incentives offered, flexible policies.

Examiner's tip

NICs or newly industrialising countries are countries whose economies are growing as a consequence of economic growth based upon industrialisation, e.g. China, Singapore, Brazil, India, South Korea and Taiwan.

(b) The key here is to **evaluate** only the **socio-economic impacts** and thus reach some sort of decision. This is a very common style of question and you should have a thorough response prepared, which should include some of the following ideas:
• **social**: urbanisation, worker exploitation, health and safety issues, cultural erosion, foreign management
• **economic**: leakage, reduction in agricultural productivity, wage disparities.

Question with model answers

C grade candidate – mark scored 8/15

 For help see Revise A2 Study Guide pp. 68–83

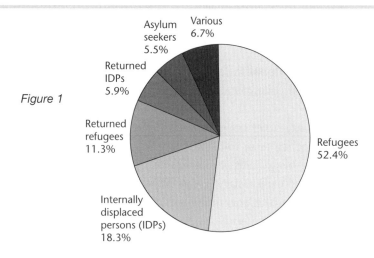

Figure 1

Asylum seekers 5.5%
Various 6.7%
Returned IDPs 5.9%
Returned refugees 11.3%
Refugees 52.4%
Internally displaced persons (IDPs) 18.3%

Study Figure 1, which shows the total population of concern to the UNHCR (United Nations High Commission for Refugees), 31 December 1999 (total 22.3 million people).

(a) Explain the differences between the different categories shown in Figure 1. **[5]**

(b) Using at least two examples, contrast the experiences of refugees and internally displaced people. **[10]**

(a) The majority of the people shown on the graph are refugees, which means that they are running away from their home country because they could be persecuted there ✔*, like those people who left Rwanda after the ethnic cleansing that took place there* ✔*. Internally displaced people are people that have just moved internally within their countries like those people who have moved because of the trouble in Colombia. The other categories are variations on these themes, but the situations are the same.*

(b) The experiences of refugees and internally displaced people are similar in many respects, and it is usually only a legal difference that means that they are perceived to be different ✔*. In Colombia there is a great deal of trouble being created by the conflict going on between the Army and the Guerrillas. The Army represent the state and the Guerrilla used to be fighting for the people, now they are heavily involved in drugs production and manufacturing. They therefore have money to buy guns and ammunition and can keep control of the population in their areas by threatening them. People who do not want to grow the drugs have no choice but to move or the angry guerrilla forces might kill them* ✔*. Most of these people become displaced people as they are forced to leave their homes but most do not make it out of Colombia as the landscape is very difficult to cross* ✔*, and they would have problems living in any of the neighbouring countries* ✔*. Many end up living on the streets and in the shanty towns of the big cities such as Cali* ✔*, where they experience many of the problems faced by refugees. They have little money and few possessions, and there is no welfare state to support them at the time when they most need it.*

Examiner's Commentary

Excellent case-study support.

Answer needs to explain, not just describe the differences.

Excellent context, but what areas in Colombia is the candidate referring to?

Two case studies provide contrast but do little for the balance of the question. A range of different experiences helps to highlight differences.

High level understanding but each of these differences needs to be developed and compared with those suffered by the Colombian displaced people.

C grade candidate continued

For help see Revise A2 Study Guide pp. 68–83

Examiner's Commentary

Refugees like those who left Rwanda were able to travel across the border during the problems, but have been able to return to their homes since. They were supported by aid agencies at the border camps on the plains of Goa ✔, *and got the international attention that meant that their problems are now being solved.*

Conclusion needed to tie all of the ideas together.

A grade candidate – mark scored 12/15

(a) A refugee is a person who has left their country of origin due to a well-founded fear of persecution ✔, *whereas an IDP may be facing the same fear of persecution but has not left the country* ✔. *The returned categories for both merely mean that these people have now been voluntarily or forcibly returned to their homes* ✔, *and the asylum seekers are those who are waiting for their applications for asylum to be granted* ✔.

Technically excellent definitions.

Improved here with reference to examples.

(b) Many recent conflicts that have forced people to leave their homes have moved people only over newly drawn up internal borders such as those that were created when civil war erupted in the former Yugoslavia during the 1990s ✔. *Many different ethnic groups have been displaced into neighbouring nations that were formerly neighbouring states* ✔. *Such forced displacement and intimidation of groups (ethnic cleansing) such as the Serbs who are currently residing in Albania, has instilled in many people a fear of persecution that means that they will never be able to return to their homes and may have lost all of their property and possessions* ✔. *The Balkans is typical of the type of conflict that has created so many refugees over the past few decades.*

Clear introduction, using details from the graph and stating the main arguments.

Good terminology needs to be incorporated into the text, not put into brackets.

The situation that erupted in East Timor during 1999 was quite different, both in cause and consequence. From a population of 800,000 some 500,000 people were internally displaced ✔, *half within East Timor and the other half in West Timor. The problems began when the country voted in favour of independence in a UN-organised referendum. Soon after the Indonesian military and anti-independence militia moved in and systematically terrorised the population of the area* ✔. *Many people returned home as the UN moved in to secure the area, but many people were left in refugee camps starved of food and water due to the actions of the militia and the particularly heavy monsoon rains that year* ✔. *The intimidation continues and many people have still to return to their homes.*

Excellent use of data to enable the examiner to appreciate not just the level of understanding but also knowledge.

The nature of the conflict determines the long-term consequences for both the refugees and internally displaced people ✔, *both fear the same type of persecution. However, refugees may have to start again in a foreign country not only without possessions or property, but also without a language or an identity* ✔.

Clear conclusion which ties together the main argument, but suggests that there is another viewpoint.

Exam practice questions

Structured question

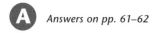
Answers on pp. 61–62

(1) **(a)** Suggest three factors that might influence your perception of refugees. **[6]**

Study Figure 2, which appeared in an article by Jane Moore in *The Sun* on Thursday, 16 March 2000.

'This week, refugee Halima Kinewa said: "We were told England was a kind country".
It turns out we are being "kind" to the tune of £32,000 a year to fund her Algerian husband Mohammed Kinewa, his two wives and their 15 children. Mr Kinewa speaks no English so is, unsurprisingly, jobless.
He has been given two fully furnished council houses for his separate families – bigamous in this country – and the children are being educated here.
Nice shirk if you can get it.'

Figure 2   *Source*: News International website, *The Sun* newspaper

(b) (i) How does the article promote negative feelings towards refugees? **[4]**
(ii) Explain the obligation of the host country to those individuals claiming asylum status. **[5]**

(2) Study Figures 3 and 4, which formed part of the 'Respect' campaign by the United Nations High Commissioner for Refugees (UNHCR) to tell people that refugees bring many contributions to their host country and therefore deserve our respect.

Figure 3: Photo of former US Secretary of State Madeleine Albright, who fled Czechoslovakia at the age of 11.

Figure 4: Photo of Canadian rapper Keinaan, who was forced to flee Somalia as a result of the civil war.

Source: UNHCR website (www.unhcr.ch)

In 2001 UNHCR celebrated its 50th anniversary. During this time it helped more than 50 million refugees. Hence the Respect campaign slogan: '50 million refugees, 50 million success stories'.

(a) What are the costs and benefits refugees bring to the country where they seek asylum? **[10]**

(b) What measures can a host country adopt to make the experience of refugees a more positive one? **[10]**

Synoptic question

(3) Study Figure 5, which shows a shanty town on the periphery of Manila, the Philippines. With reference to cities in LEDCs:

(a) Suggest what problems a new migrant might face on arrival in a large city.   **[10]**

(b) Assess the solutions that could be employed to reduce the rate of urbanisation and thus reduce the problems created by it.   **[15]**

Figure 5   *Source:* PA Photos

Answers

(1) (a)   You could use any aspect of behavioural geography here, but make sure to give a balanced explanation:
education; age; lifecycle, e.g. a person with a child of their own might have more sympathy for the plight of large numbers of parentless child refugees as were created during the Second World War; past experiences; religion; nationality, e.g. a Colombian national might have more sympathy to the plight of a Colombian refugee as they might have an insight into their problems.

> ### Examiner's tip
>
> *When you have text as a written source and you are asked to comment upon it, you can use direct quotes from it to highlight points you want. Use inverted commas and make sure that you add to the quote by assessing its validity rather than using it as a descriptive source.*

(b)(i) The article plays on the public perception of refugees:
- as **takers rather than givers** – *'it turns out we were being kind to the tune of £32,000 a year'* and *'he has been given two fully furnished council houses'*
- that **they give little to the host country** and are in some way **lazy** for not doing so *'Mr Kinewa speaks no English so is, unsurprisingly, jobless'*
- and that they have **questionable moral and cultural values** – *'his separate families – bigamous in this country'.*

(ii) The UNHCR distinguishes five types of assistance that you should consider:
emergency assistance – basic needs on arrival when still under some threat; **care and maintenance** – basic needs when not under threat: water, food, medical services, education, shelter; **voluntary repatriation** – transport, reception and assistance to help reintegration; **local settlement assistance** – for those who cannot return home, help in becoming self-sufficient; **resettlement** – longer-term facilities for permanent settlement.

Answers

(2) (a) Use the examples you have been given in the source if you know who they were, and remember that you can utilise the positive aspects that any immigrant brings with them to a country. These include: attitudes, values, beliefs, economic characteristics, social organisation, personal skills and talents, music, food, language and international understanding. To extend this answer you could discuss the society into which the refugee is placed. Clearly some are more liberal and open to such new ideas than others. Also consider the negative aspects.

(b) Consider a range of measures employed in different countries from: **migration policy** – official and unofficial, voluntary repatriation policy, asylum claim procedure, offers to resettle refugees and make facilities open to them; **welfare net** – offers of housing, education, food, clothing, help finding jobs, reception centres, assimilation, education and minimising conflict, reunification of families and support communities; **legal representation or aid**.

(3) (a) This is a wide-ranging question and requires you to structure your answer: **housing problems** – having to deal with temporary accommodation such as that found in shanty towns, e.g. bustees of Calcutta, India (legal problems over land, confrontation with authorities and police) or rented accommodation in the city centre that uses up money resources, e.g. apartments in the centre of Mexico City, Mexico; **employment** – usually informal sector employment, e.g. prostitutes in Manila, the Philippines (harsh conditions, irregular wage, potential exploitation); **friends and family** – very vulnerable when first arrived, dependant upon welfare net, familial and governmental; **crime** – the social disarray of the reception areas can often lead to crime, alienation and social stress, e.g. high crime rates in the Morros of Rio de Janeiro, Brazil; **disease** – vulnerable also to disease and poor sanitation. Try to balance these problems with some advantages of their move, or the positive impact that they can have on the city itself.

Examiner's tip

This question is contentious as it is not necessarily true that by reducing the rate of urbanisation, you reduce the problems created by it. Discuss such issues in your answer whilst covering all of the areas.

(b) Consider a range of case studies that balances reducing the migration stream by improving rural conditions and reducing the migration itself. Include ideas from: **development**, e.g. growth poles in Brazil to stimulate development in more rural areas; **transport**, e.g. develop rural transport systems to reduce isolation and enable goods to be taken to market further away; **rural/agricultural development**, e.g. irrigation schemes to increase yields; **alternative/intermediate technology**, e.g. develop better accommodation to improve rural quality of life; **exploiting local resources**, e.g. local management and exploitation of a resource; **government policy**, e.g. Indonesian transmigration where people are relocated from densely populated areas to new agricultural areas opened up by the government.
This has to be balanced with the fact that some developments in agriculture and education might actually encourage rural depopulation and not slow down the rate of urbanisation.

Question with model answers

C grade candidate – mark scored 12/20

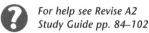

For help see Revise A2
Study Guide pp. 84–102

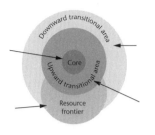

Figure 1 shows Friedmann's model of regional development.

(a) Annotate the diagram for a country or region of your choice. **[5]**

(b) Discuss the idea that economic development is principally determined by location in relation to the core. **[15]**

Figure 1: Friedmann's model of economic development

Examiner's Commentary

(a) British Isles

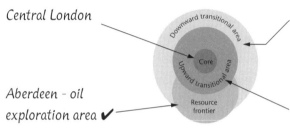

Central London

Great Yarmouth - old seaside resort/port ✔

Aberdeen - oil exploration area ✔

Reading and areas surrounding London

Clear located labels.

Some annotation and therefore explanation of why this area is the resource frontier.

(b) A location within or away from the core of a country can have a profound effect on its degree of economic development. These effects can be of a positive or negative nature leading to accelerated economic development in the area concerned ✔. Positioning in or around the given location in relation to the core can fluctuate from an area of positive economic development to an area of economic decline due to a number of outside influencing factors ✔. The city of Caracas in Venezuela can be taken as an example of this.

Once oil was extracted from the area, Caracas emerged as the single national core ✔ with people migrating towards this epicentre from the periphery ✔. The resultant cumulative growth process meant a boom in economic development for the core and a dramatic downturn in the economic development of the original areas ✔.

As the core became more attractive for business investment and foreign immigration swelled its numbers ✔, this combined with the current benefits of the oil industry, led to the development of economic sub-cores around the city ✔. This was particularly noted in the adjacent basin into which industry expanded, not possible in the physically constrained area that Caracas has grown within ✔. The establishment of industrial growth in the periphery was a governmental response to the beginnings of a relative economic decline in relation to the core periphery ✔.

In this way we can see that the location around an area's core experiences direct economic benefits especially in the periphery and local resource frontiers ✔. In contrast downwardly transitional areas often emerged in the area around the core suffering from an economic downturn due to out-migration due to the pull of the flourishing core.

Needs a time context – when did this happen?

Good process detail and use of the details of the model.

Try to avoid using terms not used in the model, such as sub-core, as you are not strictly applying the model.

Make sure to use place names wherever possible to develop a sense of place.

Question with model answers

A grade candidate – mark scored 16/20

 For help see Revise A2 Study Guide pp. 84–102

(a) British Isles
Greater London:
economic/financial
core in the City ✔,
surrounding area
service centre ✔
North Eastern
Scotland centred
around Aberdeen ✔ and the
North Sea oil fields

North Eastern England:
deindustrialised former
industrial core, loss of
shipbuilding, iron and steel,
as well as coal mining up
to 1990 ✔
M4/11 corridor with
expansion of high-tech
industries as well as ✔
associated services

Excellent located detail, which fully explains the reasons behind the choice.

(b) Nigeria is an LEDC with wide spatial inequalities of development. These are due in no small part to the development of the oil industry, and have had the result of exacerbating the already existing inequalities in wealth distribution, power and development ✔. Due to the colonial influence of the British, the core grew up around the port of Lagos ✔, where most of the primary goods that formed the basis of the relationship between colony and colonial power were exported ✔. The two largest urban areas within Nigeria are in this south-western corner of the country in Lagos and Ibadan, and this is where most of the commercial and industrial activity is concentrated ✔.

Excellent, high-level appreciation of the theory that underpins the model.

Soon after independence from Britain in 1960 oil was discovered in the Niger delta region, which went from a downwardly transitional area to being the resource frontier ✔, independent of the core area. The oil money has served to develop Port Harcourt, although most of the revenues were used in developing the poor communications and in transforming Lagos into a modern metropolis ✔. This has shifted the focus even more onto the South West and economically isolated the politically strong Muslim northern areas ✔.

A sketch diagram to show the country in simplified detail would be good here.

The clearest indication that the Nigerian government recognised this disparity and relationship between the core and periphery was the decision to construct the new federal capital of Abuja in the geographical centre of the country ✔. It hoped not only to unite the divided nation, but also to act as a growth pole that, as Myrdal suggested in his model, would promote self-sustaining economic growth ✔. This plan has not created a process of cumulative causation and the disparities grow daily thus emphasising the core peripheral relationship ✔, despite the fall in global oil prices and the reduction in revenues from it.

Good reference to another development model.

Thus until economic development is more evenly distributed, and the economy and infrastructure are in place to support it, the peripheral areas of Nigeria will remain just that ✔.

Excellent perspective on the future as a concluding remark.

Examiner's Commentary

Exam practice questions

Structured questions

A *Answers on pp. 66–67*

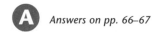

(1) **(a)** Explain what the HDI or Human Development Index is. **[3]**

Spearman rank correlation coefficient (rs) = 0.915

Degrees of freedom = n −1	Critical values for the 0.05 significance level
8	0.643
9	0.600
10	0.564
11	0.523

Figure 2

Country	HDI	GNP pc $US
Austria	0.925	28110
Bangladesh	0.364	260
Colombia	0.836	2140
Ethiopia	0.227	100
Indonesia	0.637	1080
Jamaica	0.721	1600
Nepal	0.343	210
Norway	0.932	34510
Singapore	0.878	30550
USA	0.937	28020

(b) **(i)** A Spearman rank correlation was calculated for the HDI and GNP pc, the result is shown above. Comment on the significance of this result. **[3]**

(ii) Account for the relationship between GNP pc and the HDI. **[4]**

(c) Jubilee 2000 believes that the first step in enabling less economically developed and highly indebted countries to develop is to cancel their external debt. Evaluate the options open to an LEDC government faced with a substantial debt that wishes to improve the health of its people. **[10]**

JUBILEE 2000 COALITION
a debt-free start for a billion people

(2) Study Figure 3, which shows the direction of change of international trade of the member states of the European Union.

(a) Describe the pattern of trade shown in the diagram. **[4]**

(b) Account for the relative importance of the trade with advanced industrial countries compared with less economically developed countries. **[4]**

(c) Outline the benefits to a less economically developed country of increasing its trade with EU countries. **[5]**

(d) Comment on the view that economic development always leads to an increase in human welfare. **[7]**

[AEB 2000]

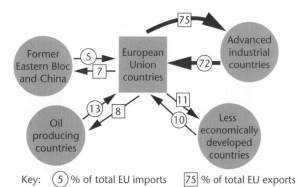

Key: ⑤ % of total EU imports   75 % of total EU exports

Figure 3: Source: Witherick, M, et al, *Environment and People* (Stanley Thornes, 1995), © Nelson Thornes

(3) Using one named country:

(a) Outline the physical constraints and opportunities for development. **[10]**

(b) Assess the strategies employed by decision-makers to reduce the spatial inequalities within this country. **[15]**

Synoptic question

(4) The UN believes that the greatest cause of the global inequalities in development is differential population growth. Examine this claim using examples of countries at different levels of economic development. **[25]**

Answers

You are likely to get at least one statistical test in any exam series, but you will usually only have to calculate the significance of the statistic using the significance tables. The chi-squared and Spearman rank statistics are the two principal ones.

(1) (a) Only two marks available, so use a clear and concise definition such as: a quality of life index that uses factors such as life expectancy, adult literacy rate and per capita income as a marker of living standards to determine a comparative index score based upon rankings.

(b)(i) Degrees of freedom = the number of paired variables − 1 = 9, therefore rs must exceed 0.600 to be 95% sure of the significance of the relationship. The basic relationship is that the higher the GNP pc, the higher the HDI; the HDI is dependent upon the GNP. Stress this relationship using some data from the table.

(ii) Countries with a higher GNP are able to spend more on health and education and this has a knock-on effect on improving life expectancy, literacy levels and living standards – hence the improved HDI.

(c) Realistically you could suggest any one of the following options:
dealing with the debt – collective default, debt for equity/nature swaps, repayment rescheduling; **development** – sensible investment of money in healthcare, reduction in military expenditure, accepting technological and monetary aid specifically designed to improve the health of the community. Cut other projects, such as road building. Use schools as platform for teaching about health issues. It is likely to be a combination of these two aspects, as once the money becomes available, it may not necessarily be used productively.

(2) (a) Make sure to stress the dominance of trade, with advanced industrial countries having a trade surplus with all but the oil producing countries. Use details and data from the diagram to support your answer.

(b) This can be due to a number of factors:
- high value of manufactured goods traded with the advanced industrial countries
- importance of trade in manufactured goods
- limited demand of any high-value goods except for oil from LEDCs
- limited trade opportunities with the former Eastern bloc.
For four marks you should make sure that you balance your depth and breadth, not too deep, but sufficiently broad to cover at least three of the points.

(c) The issue here is the nature of the relationship between the trading characteristics and the level of economic development, e.g. less economically developed – dependence on primary goods – low prices/vulnerable to price changes/could be damaged by natural hazards. Beyond this, you should concentrate on the **multiplier effect**, e.g. capital to invest, attracting foreign investment.

*You should know a wide range of indicators of development so that you can use them in questions such as the one below. You should use the simple **quantitative** (things that can be measured) indicators, e.g. economic, social and demographic, as well as the **compound** (made up of a number of different indicators) indicators, e.g. **PQLI** (Physical Quality of Life Index), **MEW** (Measure of Economic Welfare), **PPP** (Purchasing Power Parity) and **HDI** (Human Development Index).*

Answers

(d) You should look for a balanced argument here. You could argue that economic development does lead to increases in human welfare and back this up with specific reference to qualitative indicators and specific countries. Clearly the contrary can also be argued, and you should try to incorporate the ideas of inequality, poverty, cultural impacts and environmental degradation. You might conclude by suggesting that using compound indicators might give you a clearer picture instead of focusing on debatable details.

Examiner's tip

It is important in development studies to have a holistic understanding of at least one country and what underlies its current state of development. Areas to consider might be: physical, colonial/neo colonial, TNCs, trading blocs, population growth, diseases, education, debt, loans, government action.

(3) (a) There are a number of ways to address this question but remember to use the flexibility offered to you by the title. You can choose a country in any state of development, anywhere in the world, but you should make sure that you know the country sufficiently well to put forward a balanced viewpoint. Physical aspects could either hinder or aid development, but they will certainly lead to spatial variations in development:
- **resources** – natural vegetation, soils, climate, minerals, water supplies
- **topography and relief** – river valleys, mountain ranges, altitude.

(b) You should have already made reference to some of the spatial inequalities in your chosen country, so this second part should flow from that. The simplest way to look at development policies on a national scale is to consider that they could be either:
1. **'top-down'** – state or government makes decisions about development, usually without consultation with the people and focuses upon urban areas, e.g. **growth poles** in Brazil, **new capital** of Abuja in Nigeria, **corridors of development** around London in the UK, **motorway construction** in Portugal with EU financial assistance, the Aswan **dam construction** in Egypt, or
2. **'bottom-up'** – local people make decisions and manage their own development, usually focussed on rural areas, e.g. **micro-hydro** projects in Nepal, **embankment building** in Bangladesh to prevent flooding, **community schooling** in poor barrios in Lima in Peru, **ecotourism** in Costa Rica.

MEDCs will tend to be more top-down and LEDCs bottom-up, but don't be afraid to use specific locations when assessing bigger policies.

(4) This essay focuses you on **population** but as it is synoptic you need to consider the wider picture and include ideas from development, industry, tourism, agriculture and even weather and climate and environmental hazards. Considerations for high population growth might include:

capital generated merely supports population growth; lack of employment opportunities puts pressure on resources; surplus labour and limited opportunities creates migration and therefore promotes unequal development; making a growing population into a resource.

You might include a whole range of other ideas for balance: TNCs, industrialisation, politics, trade, neo-colonialism, colonialism, capital, government policies and natural hazards. Conclusions could include extension ideas such as pro-/anti-natalist policies and other ways to alter population growth.

Question with model answers

 For help see Revise A2 Study Guide pp. 104–124

Center Parcs is a holiday company with three purpose-built leisure complexes based on the concept of a recreational theme park in a woodland environment. The parks are all car-free and the accommodation has been built around a central climate-controlled dome. Study Figure 1, which shows extracts from a Center Parcs press release, and a photograph from one of their Villages.

Examiner's Commentary

Center Parcs is the winner of the Landscape Institute 1999 Award for Landscape Management. This award recognises Center Parcs ongoing commitment to creating outstanding natural environments at each of its three UK Villages: Sherwood Forest, Nottinghamshire; Elveden

Forest, Suffolk; and Longleat Forest, Wiltshire. The Center Parcs Forest Management Plan was hailed by the Landscape Institute as a landscape management plan 'that is a fully co-ordinated, integrated, dynamic, enlightened and workable document'.

Each Village has a variety of glades, waterways and woodlands ranging from mature coniferous plantations, replanted coniferous and ancient deciduous woodland. Small lakes and streams have been introduced to enrich the ecological diversity. To attain the diversity of habitats, landscape variety and spatial character, each Village has a ten-year Forest Management Plan. This is complemented by annual ecological monitoring studies. Center Parcs also works to its own Biodiversity Action Plan, which highlights targets for each Village, taking into account the local targets of species recovery set by English Nature.

Figure 1 Source: Center Parcs press release

(a) Evaluate the costs and benefits that arise from the management plan outlined in Figure 1.   **[7]**

(b) Account for the increasing number of visitors attracted to such theme parks in MEDCs.   **[13]**

C grade candidate – mark scored 11/20

For help see Revise A2 Study Guide pp. 104–124

Examiner's Commentary

(a) There must be good reasons for a large leisure company to undertake such a large financial investment when it does not have to do so. Doing all of these positive things to the woodland environment will help it to be seen in a very good light by potential customers ✔ and hopefully attract many visitors to the parks ✔. It may also sustain the environment into the future ✔.

You cannot assume that the examiner can read your mind. What are the good reasons?

Too short to evaluate the article.

(b) The reason that theme parks are becoming so successful is that many of them are copying the example set in the USA where niche marketing has meant that everybody in any family is catered for, from the small baby to the most elderly pensioner ✔. In Britain Legoland caters for young families with children, but Chessington World of Adventures caters for older children and their friends, with or without their families ✔. Euro Disney caters for those families with a little more disposable income ✔, and Center Parcs for those people who want to get away from it all and have a break in the country ✔.

You only need a couple of examples to highlight your ideas.

Each of these attractions has to be well advertised and marketed, and as with many of the ideas for the different parks, much of this information comes from the USA where parks like Disneyland and Disney World have been open and operating for nearly 50 years ✔. This has enabled them to develop a specialist niche in the market that European parks are only just beginning to take advantage of.

Essentially a repeated point, with reference to the USA.

Clearly people in MEDCs have the advantage of increasing wages ✔ and increasing numbers of cars ✔ which enable them to get the such parks more easily, spend more time there and visit more frequently ✔.

Points which need expansion and to be addressed using technical language, i.e. disposable incomes.

Question with model answers

A grade candidate – mark scored 18/20

 For help see Revise A2 Study Guide pp. 104–124

Examiner's Commentary

(a) The clear outcome of the management plan for Center Parcs has to be based upon sound financial considerations as they are clearly a commercial company and need to make a profit ✔. The environmental benefits are clear in that they are doing much work to plant and reintroduce rare species into the woodland environment ✔, so encouraging biodiversity and preserving different natural environments. The management plan also includes monitoring and goal setting, all of which can only be beneficial for the environment ✔. The costs come from the money that needs to be invested by the company to carry out and maintain this extensive management programme ✔, and any damage that the construction of the actual building might cause to the natural environment ✔. This all creates a very positive environmental image for the company in our increasingly environmentally aware society ✔, and might help to reduce the costs of advertising by creating its own publicity ✔. The benefits clearly outweigh the costs in both environmental and economic terms, as long as the plan is kept going.

Probably too much detail for a seven-mark question, excellent coverage but too long in the time given.

Excellent understanding of processes.

(b) Center Parcs like many other theme parks in the UK such as Alton Towers and Legoland, are so successful in part because they are well advertised ✔, but that does not fully explain why people keep coming back.

In most MEDCs there is a growing proportion of middle-class people who are looking for more regular, short-term breaks and days out ✔. Such places provide them with this opportunity and the owners of such parks know what people want when they are there. Alton Towers has a wide range of family-friendly rides and attractions to keep the whole family entertained for a whole day ✔, and although many of the parks are in themselves expensive, in the new millennium people are beginning to spend more of their disposable income on leisure and recreation ✔.

Also increasing car ownership ✔ and the reduction in the relative price of public transport ✔ has increased people's personal mobility in the UK and enabled them to get to such parks over a weekend ✔. The most successful example of this is Euro Disney near to Paris, which after a poor opening, is now one of the top tourist destinations in the whole of Europe ✔. It also offers that little piece of the USA on the doorsteps of most Europeans ✔, with the high standards of hygiene and safety that all responsible families would look for in a weekend break ✔.

Some statistics would be good to include here, i.e. UK car ownership figures for 1980 and 2000.

Ultimately there is a demand for such parks as well as an increasing number of them being built ✔, and while the standards remain high we will continue to visit them in increasing numbers.

Excellent clear conclusion which relates back to the question.

Exam practice questions

Structured question

Answers on p. 72

(1) Study Figure 2, which shows a Yak herders' hut used for ecotourism beneath Jaljale Himal in Nepal.

(a) Using the photograph, distinguish between primary and secondary resources for recreation and tourism. **[4]**

Figure 2

Source: Steve Conlon

Six main themes important in policies for sustainable tourism:
1. Planning and control of the spatial distribution and the character of tourism developments
2. Surveys of resource evaluation and impact assessment before development takes place
3. Integration of tourism into other aspects of regional planning
4. Increasing involvement and control by local and regional communities
5. Identification of the type of tourism appropriate to the resources and environment
6. Establishment of a carrying capacity that balances conservation and development values

Source: *Leisure, Recreation and Tourism*, Prosser, R, Collins Educational, 2000

Figure 3

(b) Using Figure 3, select three of the themes and explain the advantages of their application in wilderness regions such as that of Nepal. **[6]**

(c) How could mass tourism become more sustainable in the light of the guidelines laid down by ecotourism? **[10]**

Extended question

(2) **(a)** Evaluate the short-term impacts for a city of hosting a major sporting event such as the Olympics. **[17]**

Read the text in Figure 4 below, which shows the objectives for the Athens 2004 Olympics.

(b) How will the aims stated in Figure 4 help the long-term development of tourism in a large city like Athens in an MEDC? **[8]**

Figure 4　　　　　　　　　　　　　　　　　　　　　　　*Source*: Athens 2004 Official Olympic website

---

Athens 2004 Olympic objectives
- To organise the best Olympic Games so that Athens becomes a point of reference for the future
- To reposition the Olympic Ideals in the contemporary frame of the 3rd millennium – the Games of the future in relation to the traditional Games
- To develop and establish the Cultural Olympics
- To establish and promote the institution of the Olympic Truce
- To contribute to the improvement of the environment and the quality of life
- To contribute to the rapid and stable development of the country
- To successfully balance the Olympic Ideals with the Games' commercial aspect
- To extend and promote traditional Greek hospitality
- To leave future generations a legacy.

Answers

(1) (a) **Primary resources** e.g. scenery, climate, ecology, history and heritage. Use examples from the picture and remember that as the herder's hut is used for ecotourism, the **secondary resources** (those used for tourism) e.g. accommodation, catering and entertainment, may not have been specifically constructed for tourism, but adapted for the purpose.

(b) Use information from the picture to help you. It is clearly a remote, unspoilt wilderness but sustainable tourism suggests that development will take place, e.g. integration of tourism into other aspects of regional planning – transport and communications could be developed to allow the tourists easier access as well as providing the local people with better routes to market for their goods and services.

(c) Focus on sustainable tourism here (managed use that does not cause any long-term environmental damage). Incorporate case studies in these longer questions to develop your ideas. Practical suggestions could include:
accommodation; using insulation to reduce heating costs, using local materials in the construction of facilities, incorporating sustainable energy use into buildings; **transport**; promoting environmentally friendly transport such as bikes, walking, boats, reducing the need for cars; **environment**; maintaining and promoting biodiversity and conservation, installing boardwalks, understanding the **carrying capacity** of the environment; **catering**; using local produce, choosing washable table linen, reducing packaging materials; **host/visitor relations**; tourist education, information boards, promoting better working relationships with locals; **financial**; using local management, using local financing/capital so as to reduce leakage.

(2) (a) With reference to specific cities, consider the following ideas, in the **short term** (before, during and shortly after the event):
economic, e.g. direct/indirect employment created, investment in the area, tourist spending stimulates the local economy, advertising for the local area and country, global prestige, local prices rise, deflection of funds away from other areas of need; **social**, e.g. pace of life increases, local people either adapt or withdraw, increased provision of sports and housing facilities, different social interactions due to mix of peoples attending, increased sense of national pride and social cohesion; **environmental**, e.g. increased air and noise pollution, strain on resources, increased traffic.
Evaluate the impacts and reach some conclusion, e.g. Sydney 2000 was considered to be an unqualified success, whereas Atlanta 1996 was not, due to poor management and organisation; corporate influences also tarnished the image of the games and the city. Include positive and negative impacts.

Examiner's tip

Always make sure that you read essay questions twice before you attempt to plan them. At first glance, this question might appear to require you to discuss Athens, but you could use any MEDC city that you have knowledge of. You might do better to discuss a city that you know well, such as a city you live in; local knowledge in essay questions can yield excellent results.

(b) The focus is on the long term development, you should look at how these aims might help achieve this, the first wishes to establish Athens as a world tourism destination as well as creating positive images of the city through the efficient and well-organised management of the games. Consider the cultural implications e.g. establishment of art galleries and parks; improved infrastructure and public transport; attraction of global companies (e.g. improved shopping facilities).

Question with model answers

C grade candidate – mark scored 6/10

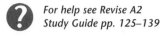 *For help see Revise A2 Study Guide pp. 125–139*

'Sustainable agriculture can be defined as the ability of an agricultural system to be reproduced in the future without unacceptable pollution, depletion or physical destruction of natural resources...'

Source: Bowler, I, *Agricultural Change in Developed Countries*, Cambridge University Press, 1996

(a) Outline the threats that modern agricultural methods practised in MEDCs pose to natural resources. **[10]**

Examiner's Commentary

(a) There have been a great number of recent developments in agriculture from GM foods to the use of pesticides. They have appeared in many of the local and national newspapers and have caught the attention of the population to such a degree that we have decided to change what we buy in the supermarkets and grocery shops of our country. Many people do not know just what these recent developments mean for the countryside and are frightened by scare stories ✔.

Good, interesting introduction; it needs to define the terms used in the question and stick to the point.

Genetically modified foods have damaged many other crops in areas next to where they have been grown, and many people worry about their impact upon the people who eat GM food ✔. They are not grown any differently from other non-GM crops so they do not do any extra damage in that way, but there are other techniques that damage the natural resources.

With current events, make sure that you know the situation perfectly.

Large machinery can compact the soil and damage the structure and drainage ✔, they can also remove vegetation so efficiently that there is none left to protect the soil against the wind and rain ✔. Such big machines also require large turning circles and therefore many hedgerows have been removed to make more efficient use of space ✔.

Excellent understanding of process – but some expansion of the wider impact.

The real problems with the natural environment may be in LECDs where population pressure is causing farmers to use higher yields from increasingly small plots, and thus causing them to harm their natural resources ✔.

Good wider reference to both the MEDC and LEDCs, but does it add to or conclude the question?

Question with model answers

A grade candidate – mark scored 9/10

 For help see *Revise A2 Study Guide pp. 125–137*

Examiner's Commentary

(a) There are many aspects of modern agricultural methods that cover a wide range of sciences. The mechanisation of many of the manual jobs has helped make the sowing, tending and harvesting of crops ✔, as well as the husbandry of animals, far easier ✔. Chemicals have been used to encourage the growth of crops and animals as well as to discourage pests and vectors ✔, and recent developments in genetic modification have developed crops beyond what we could have expected from simple cross-breeding ✔. All of these practices impact upon the natural resources that a farmer has available to them, from the soil and soil water, to wider impacts on wildlife and drinking water.

This is a very thorough introduction, but is it interesting? Try to get the examiner's attention by introducing interesting ideas.

The first aspect to consider is the pollution of the water and soil. The use of pesticides and fertilizers in large quantities can leave poisonous residues in the soil and water, building up in plants and animals alike ✔. They can also leach into ground water and cause the eutrophication of surface water ✔. These are not the most direct of threats for the farmer, as they may not impact upon production. Genetically modified crops may yet prove to have harmful impacts in the long run, but their pollen may 'pollute' other crops in the long run ✔.

Good use of the source to form the structure of the piece, i.e pollution, depletion or physical destruction.

Could have made reference to hedgerow depletion and the loss of biodiversity that some modern techniques cause.

The depletion and physical destruction of the soil by overgrazing, overcultivation, deep ploughing and poor soil conservation techniques is of great concern, as all of these techniques damage the fragile soil structure ✔. It is usually the use of heavy machinery such as combine harvesters and tractors with large ploughs that leaves the soil compacted or exposed and vulnerable to erosion.

Be specific about the type of erosion – wind and water (or pollution – water, air, land, visual).

Thus it can be seen that modern agricultural techniques damage our natural resources in a number of long- and short-term ways. In the long term, however, even human beings are natural resources and thus we must be considered when trying to solve the problems that such modern agricultural techniques cause ✔.

Good linkage into the second half of the essay.

Exam practice questions

Structured questions

A *Answers on pp. 76–77*

(1) Figure 1 below shows the 10 countries with the greatest dietary energy deficit per head of population. The deficit is expressed in kilocalories per person per day (2,000 kilocalories being the minimum recommended daily intake of calories).

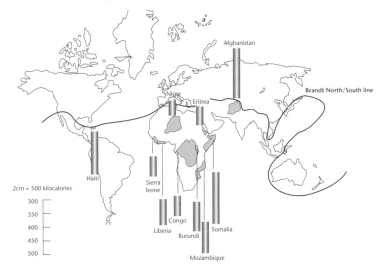

Figure 1

(a) Describe the pattern of greatest dietary deficit. **[3]**

(b) Outline the possible causes of such deficits. **[7]**

(c) (i) Suggest the potential solutions open to both governmental and non-governmental organisations to overcome such deficits in LEDCs. **[5]**

   (ii) From the solutions you have noted in **(i)** above, explain which should prove to be the best long-term solution. **[5]**

(2) Figure 2 below shows the growth of organic agriculture in the UK.

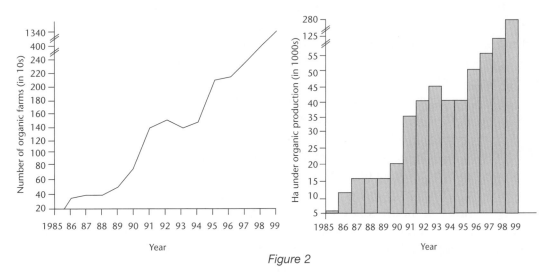

Figure 2

(a) Assess the relationship between area under organic production and numbers of organic farms in the UK. **[3]**

(b) Suggest how consumer pressure and environmental concerns may have caused the pattern of growth that appears on the graph. **[7]**

(c) Discuss the idea that the advantages of organic farming do not outweigh the problems for either the farmer or the consumer in more economically developed countries. **[10]**

Exam practice questions

Essay question

(3) (a) What factors continue to limit the expansion of food production in some parts of the less economically developed world? **[15]**

(b) How have some areas in the less economically developed world succeeded in expanding food production? **[10]**

[AQA (AEB) 2000]

Synoptic question

(4) Assess the role of agricultural change in meeting the needs of a growing global population. **[25]**

Answers

> **Examiner's tip**
>
> *At A2 level you may be given extra information on maps. This means that you have to make decisions about which piece of information is important. Here you are presented with a very simple map that has the Brandt North/South line drawn upon it and little other information, so use this to aid your description of the distribution.*

(1) (a) The countries are all south of the Brandt North/South line, principally in Central Africa, with exceptions in the Caribbean and Central Asia. Use figures from the map with as much accuracy as you can to aid your description.

(b) Consider physical factors such as drought or natural hazards such as flooding or hurricanes. Make sure that you use examples here to back up your ideas, as the question is not specific about why the countries on the map are experiencing such difficulties.

At a higher level you should really be outlining why many countries are experiencing civil war or political unrest which may mean that it is difficult to grow crops on the land, or that supply lines for vital foodstuffs are being broken by opposing forces. Better still, you should comment on the geopolitical situation in the *region* that these countries are in, which may explain why they are not able to feed themselves.

(c) Talk about: short- and long-term solutions, government and non-governmental organisations. International food aid distributed by NGOs such as Oxfam might deal with the short-term deficit, but longer-term solutions are the only real answer to solving this problem. You should make some reference to techniques that could be used to improve yields, or **intermediate technologies** that could be employed to **help people to feed themselves**. Ultimately you should try to deal with the cause of the problem, be it political arbitration or better hazard management, such as reserves of food in case of a hurricane. You must argue what you believe is the best solution.

(2) (a) It is a simple proportional relationship with the area under production growing at a rate comparable to the number of farms, but you should make reference to the different rates of growth, e.g. between 1985–1991 and 1997–2000. You could make your answer better by charting the average size of the farms to further detail this relationship, i.e. 1985 200 ha, 1992 270 ha, 1999 210 ha.

Answers

(b) Begin by discussing what consumer pressure and environmental concerns may have caused a change:

concerns over beef in the BSE scare; increasing awareness of the impacts of pesticides and fertilizers on the environment and upon humans; more green awareness after high-profile media events like the Rio summit and GM crop trials.

Make sure that you stress that such pressures and demands led to the growth in organic production to meet the needs of the new 'green' consumer and the response of supermarkets.

(c) You do not need to reach a conclusion in this question, but you should consider some of the following points and present a balanced argument:

Consumer	Farmer
+ fewer chemical residues	– long time to convert with lower yields afterwards
+ perceived healthier food	– greater losses from pests and diseases
– higher bacterial content	– more labour intensive
– more expensive produce	+ less damaging to the environment and uses less energy
	+ higher returns

Examiner's tip

Ensure that you make a simple plan for essay questions with titles that could include a very wide range of ideas. This will give you a chance to consider your structure and the appropriate case studies you should use.

(3) (a) There are a wide range of ideas to consider here, which could be roughly grouped as follows:

climatic; political; technology; education; capital; diseases and pests; land ownership. The command word is simple; i.e. 'what', so you only need to get an overview and make sure that it is balanced and exemplified.

(b) Hopefully your planning will have enabled you to see the links between parts **(a)** and **(b)** and therefore your answer to this question can neatly follow on. Ideas such as: technology – intermediate or advanced technologies such as the Green Revolution in India; education – to improve literacy; capital – higher prices for cash crops or micro-financing; diseases and pests – control diseases with immunisation, or improved water supplies; meaningful aid and population control. Use case studies.

(4) Clearly you need to take a synoptic view, and include not just agricultural ideas but also concepts from population. You could begin by outlining the current state of the world's population, and where the growth is concentrated. This will enable you to present the split between the less and more economically developed world, both in terms of population growth and ability to deal with such change. Then you can address the issues as you perceive them. You need to cover the different aspects of agricultural advancements, such as GM foods and mechanisation, as well as the things that are hindering development, like poor levels of literacy, debt and diseases. You have the ideal opportunity to include some theory here, and could use ideas from **Malthus** and **Boserup** as well as under- and overpopulation. Population control is also significant and ultimately could determine a country's ability to feed itself, e.g. Mauritius and China. Just make sure that your essay is grounded in case studies, and even better in issues, and that you reach a conclusion as the question asks you to.

Question with model answers

C grade candidate – mark scored 14/25

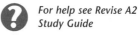
For help see Revise A2
Study Guide

Figure 1 shows the way in which a city consumes resources and produces waste.

The European Environment Agency (EEA) estimates that an average European city of one million inhabitants requires 11,500 tonnes of fossil fuels, 320,000 tonnes of water and 2,000 tonnes of food every day. It also produces 300,000 tonnes of wastewater, 25,000 tonnes of carbon dioxide and 1,600 tonnes of solid waste.

Inputs

Food

Energy – nuclear, coal, gas, wood, alternative

Manufactured goods

Water

Outputs

Land pollutants – landfill, chemical residues

Water pollutants – dumping of organic and inorganic waste and refuse

Air pollutants – CO_2, SO_2, NO_2

Source: The Photographers Library   *Figure 1*

Discuss the contention that the route to building more sustainable cities is through reducing demand for resources rather than increasing supply to meet such demand. **[25]**

Sustainable development is defined as development which meets the needs ✔ *of the present without damaging the environment for future generations, so the best way to achieve this is to reduce the demand for resources* ✔ *so as not to put pressure on the environment.*

Many MEDC cities have set about a process by which they are reducing the inputs into the city system ✔*. Cities like Sheffield have put in place an efficient tram system in order to reduce the need to use motorised transport to get to work* ✔*, and have improved the quality of the air as a consequence* ✔*. Even an LEDC city like Mexico City with its poor air quality due to the huge numbers of cars on the roads* ✔*, have introduced a restriction on driving cars based upon car licence plate numbers* ✔*. In London Thames Water has spent large sums of money upgrading the Victorian water mains* ✔ *and installing water meters in order to reduce the demand for water* ✔*.*

Other cities like Norwich have tried to put into place an efficient recycling system in order to reduce the outputs ✔*, but it is to the LEDC city that we must look in order to see efficient recycling. Cities such as Sao Paulo naturally recycle most of the waste they produce as they have an underclass to do it* ✔*, but often it is these same cities that have the worst pollution* ✔*. Mumbai has some of the worst air and water pollution due to poor environmental laws and sanitation, and this does not help the city itself to remain sustainable as it threatens the health of the entire population* ✔*.*

Although it is important to make sure that people's needs are met, it is more important that the environment is preserved for future generations. The

Good definition of terms, but better integrated into the text in an interesting way.

The candidate needs to understand the concept of 'social justice', or the equal spatial provision of services,

Good background material.

C grade candidate continued

For help see Revise A2 Study Guide

standard of living will naturally grow as it does for the population as a whole and not at the expense of the non-urban areas which serve the city. Thus we must be more efficient in our use of resources as well as reducing our demand before we can hope to meet increasing demands ✔.

Examiner's Commentary
The argument needs balance, e.g. to show that a city must often 'grow' in order to survive.

A grade candidate – mark scored 18/25

The blueprint for building sustainable cities was outlined during the Rio Earth Summit of 1992 in Agenda 21 ✔. It suggested a holistic approach to urban sustainability that takes environmental concerns seriously, but balances this with considerations of economic, social and community concerns ✔. However, if sustainability is the ability to meet the needs of the present without compromising the ability of future generations to meet their own needs ✔, then we have a problem as urban sustainability means that the whole community needs to have access to the full range of urban resources from water and sanitation to education and health care ✔. Surely this means that more needs to go into the system?

Clear use of the definition of sustainability in context.

Norwich City Council recently won an EU award for their sustainable city plan ✔. One of the most important aspects of any approach to sustainability is planning ✔, as this means that the resources that are available are evenly distributed ✔. This was at the heart of the Norwich City plan, and meant that future services such as GP surgeries, schools and hospitals will be built with the whole community in mind ✔. To achieve this, the local community was involved in the planning process ✔.

Good use of a local case study, clearly familiar to the student.

They have therefore increased the supply of such facilities as sports and leisure facilities with the redevelopment of a large area of former railway land into a large, integrated leisure complex called the Riverside ✔. The University has had a huge investment in its sports facilities, and much time has been spent providing and retaining art, cultural and historical facilities such as the 52 churches within the city ✔.

Excellent balance and structure, but what about LEDC cities?

They have attempted to reduce demand in many areas. Attempts have been made to increase the % of waste that is recycled ✔, introduce water meters to reduce water usage in this very dry area ✔, improve energy efficiency and investigate alternative methods of energy generation ✔. One of the major developments has been the introduction of an integrated transport network which hopes to link up the railway station and city centre with buses to all of the outlying areas ✔.

Good focus on one place, but specific reference to other sustainable cities and their approaches would help.

Thus urban sustainability is a careful balance between reducing demand and meeting supply ✔, and must always bear in mind that a city is an economic organism which needs to produce money for its population in order to stay alive ✔. Cities such as Norwich and further afield in the USA and Brazil have also seen the advantages of projecting this image to the rest of the world in order to attract further investment ✔.

Exam practice questions

Structured question

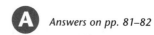

Answers on pp. 81–82

(1) Read the text shown in Figure 2.

> 'The challenge facing world agriculture today is to provide food, fibre, and industrial raw materials for billions of people – without jeopardising the future productivity of our natural resources. Meeting this challenge will require the continued support of science, research and education programmes. To meet environmental and food safety goals we also need to coordinate international food safety and environmental policies.'

Figure 2   *Source*: Edward Madigan, US Secretary of Agriculture (1991), Courtesy USDA

(a) (i) Outline one recent technological development in agricultural food production.

    (ii) Suggest two advantages of such a technique for the farmers who would employ it. **[7]**

(b) In what ways can education be used to meet the challenge facing world agriculture? **[6]**

(c) Comparing countries at different levels of economic development, analyse the decision-making behind the introduction of more sustainable agricultural practices. **[12]**

Extended question

(2) Study Figure 3, which shows steel's recycling loop.

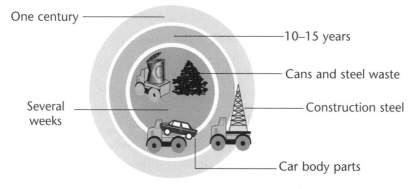

Figure 3

(a) Comment upon the feasibility of recycling steel as suggested by Figure 3. **[5]**

(b) In what ways have advanced manufacturing and distribution techniques enabled the manufacturing industry to become more efficient and thus more sustainable? **[20]**

Synoptic question

(3) Study Figure 4, which shows the consumption of various resources by the world's richest countries, which represent 20% of total global population.

(a) Suggest the possible reasons for the imbalance between population and resource consumption shown in Figure 4. **[10]**

(b) How might individuals in the world's richest countries 'think globally and act locally' in order to reduce the resources they consume? **[10]**

Answers

*The most important definition to remember is the one for **sustainability**, as it can be adapted for all sustainable situations: 'to meet the needs of the present without compromising the ability of future generations to meet their own needs'. You should be able to quote it and understand its implications for all aspects of geography.*

(1) (a) This is your opportunity to show your wider reading on agricultural research, and you could include any current topics such as genetically modified foods, but make sure to clearly explain what the technology is designed to do and what the advantages are for the farmer, e.g. the advantages of genetically modified foods for farmers are that they could maximise yields which would increase profits, and could produce pest-resistant crops thus reducing the need to apply expensive pesticides.

(b) Use the text here, as there are two different aspects that need to be covered:
1. **to provide food, fibre and industrial raw materials**, e.g. improved farming techniques such as intensive irrigation which could increase yields
2. **to meet environmental and food safety goals**, e.g. more efficient application of fertilizers which would reduce the quantities required and the quantities that leach into the groundwater.

Explain how farmers could be educated in these methods: media/poster campaign; training programmes (NGO and government) etc.

*The key to much sustainable development is that there is usually a **short-term cost** in order to achieve a **long-term benefit**.*

(c) The decisions are based upon a number of different issues, but the focus should be upon **economic gains/losses** and **environment awareness/concerns** rather than the other behavioural aspects such as age, education, gender, capital and resources, although they all affect decision-making on a personal level, e.g. the decision for a UK mixed farmer to 'go organic' may be based upon environmental concerns but has to be a sound economic decision in both the short and long term, whereas for a Pakistani subsistence farmer the adoption of more sustainable techniques may be a sound economic decision as it could preserve his resources in the long term.

(2) (a) The diagram suggests that the steel in cans is recycled several weeks after use, in cars after 10–15 years and in buildings after a century. However, you should consider the feasibility of such recycling which depends upon a number of things – the ability to recycle, awareness of recycling programs, environmental awareness, use of recyclable materials – and all of this can be dependent upon the level of development of the country.

When studying the development of modern manufacturing techniques the key is to understand the difference between 'Fordist' and 'post-Fordist' production techniques. Fordist techniques are based upon those employed by the original Ford car plants, where mass production provided uniform affordable products. Henry Ford is famous for saying that you can have any colour of car you like as long as it is black. Today you can order a car to your exact specifications, but the production process has to be highly efficient to make this process affordable.

Answers

(b) This is where you should utilise your knowledge of a manufacturing industry or company. They are bound to have experienced many changes throughout the 1980s and 1990s in order to make their production process more efficient. Use this detail to argue the case for sustainability. Include ideas such as: 'just-in-time' manufacturing, computerised stock control, continuous quality control, computerised assembly, more efficient methods of production, recycling of waste at every stage of manufacturing, increased quality and reliability of goods lengthens life and thus replacement time, and the use of higher quality recyclable components in manufactured goods. Remember that every industrial system has inputs and outputs; you are trying to reduce both inputs and waste outputs, as well as ensuring quality and hence longevity.

Examiner's tip

*Synoptic questions may be open-ended and thus leave scope for personal opinion, but an A2-level geographer should always bear in mind two things: **1. Balance** – your essays should offer a balanced perspective on any issue. **2. Support** – you should always support all of your lines of argument with case studies. Synoptic case studies can also be better if local and thus you are aware of the wider perspective.*

(3) (a) You must introduce technical language here such as LEDCs and MEDCs, and fully describe the graph so that you can then explain it. As it is a very open-ended question there are a wide range of arguments you could focus on: **development and trade**, e.g. globalisation, the 'New World Order', colonialism, trading blocs, diet; **industry**, e.g. industrial development, TNCs; **agriculture**, e.g. agricultural production.

(b) The focus here could be personal and thus you could address the sustainability issues that occur as a consequence of the life you lead. Ideas must range from **perception** and thus information and education, to **action**. Action could be taken upon:
transport; food, consumables and waste; energy requirements; management and planning; and a whole range of other areas. Be sure to use a wide range of support materials.

Examiner's tip

*Always try to consider the **environmental, social** and **economic impacts** of any concept that you study in geography, e.g. the impact of TNCs or sustainable tourism projects. Too often we concentrate upon one aspect depending upon whether we are studying human or physical geography, but synoptic concepts such as these require a more holistic view.*

Mock Exam

Time: 1 hour 30 minutes

Instructions
Answer 3 questions.

Note to the reader on the mock examinations
In the mock examinations that follow we have attempted to cover the approach adopted for assessment by as many of the specifications as is possible. This means, of course, that questions in the Practice Mocks are not necessarily of the same length or value. This should not concern you, as it is the style of questioning and the actual practising that is important!

Before the final exam, you should familiarise yourself with the assessment style and length of the examination in your specification. You should check this by looking at the Specification Grids in the AS and A2 Study Guides produced by Letts.

(1) Study the map below which shows the distribution of volcanoes around the world.

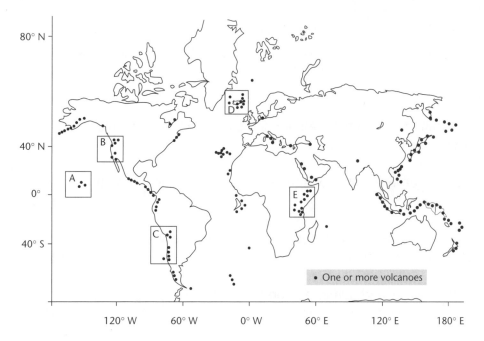

(a) (i) What is a volcano? [2]

(ii) Describe the distribution of volcanoes shown on the map. [3]

(b) (i) Using the letters on the map, identify:

    1. a constructive margin   **2.** a destructive margin. [2]

(ii) Explain why volcanoes occur at location A. [3]

(c) (i) How are intrusive landforms formed? [4]

(ii) Explain how igneous activity may provide economic benefits. [6]

[Edexcel specimen]

(2) Using earthquake case-study material, explain how the impact and management of their primary and secondary effects depend on the level of development of the country involved. **[10]**

(3) Explain why, despite the hazards in volcanic regions, such areas often experience population increase, settlement expansion and economic growth. **[25]**

[CCEA]

(4) (a) The diagram below shows a plant succession in Shropshire.

Habitat description	Reed swamp	Marsh or fen	Open wooded fen	Closed wooded fen	Woodland		
Number of species of plant	6	10	14	26	18	14	10

(i) Name the arresting factor in the diagram and state the stage at which it ceases to dominate. **[2]**

(ii) Why are there a smaller number of species of plant in stage 8 than there are in other stages? **[2]**

(iii) Describe how vegetation characteristics in the diagram show that this is an example of a plant succession. **[4]**

(b) Explain how both physical and human factors may arrest the progression of a plant succession, such as that shown in the diagram above. **[7]**

[AQA]

(5) **(a)** In the context of basin hydrology, what is meant by the following terms?

    (i) Interception. **[2]**

    (ii) Throughflow. **[2]**

(b) If you were undertaking fieldwork involving the collection of data, what method of sampling would you use to choose locations for measuring rates of interception? **[5]**

Study the diagrams below:

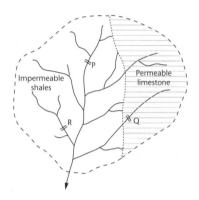

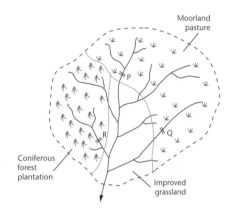

Figure 1a: Drainage and geology   *Figure 1b: Drainage and land use*

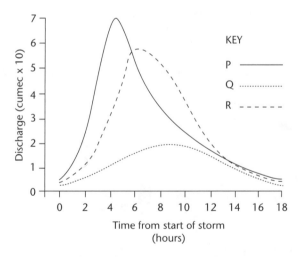

Figure 2: Storm hydrographs for each of gauging stations P, Q and R

(c) **(i)** Describe the differences between the storm hydrographs for each of stations P, Q and R. **[7]**

    (ii) Suggest reasons for the differences between hydrographs P, Q and R. **[9]**

[AQA]

(6) 'River management schemes are a mixed blessing.' Using examples describe the advantages and disadvantages of some schemes you have studied. **[20]**

(7) Study the diagram below.

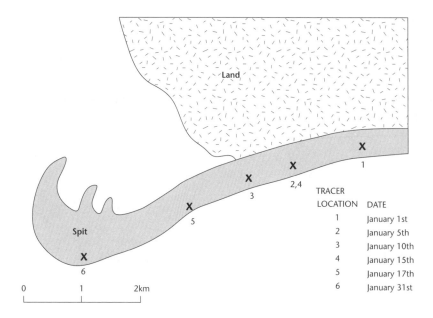

(a) (i) State the distance moved by the tracer over the month shown on the map. **[2]**

(ii) Between which two dates was the tracer moving the fastest? **[1]**

(iii) Suggest an explanation for the variations in direction and rate of movement of the tracer. **[4]**

(b) Study the diagram below, which shows wind strength and direction near the spit shown on the map above.

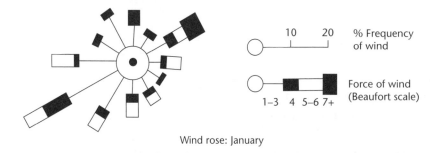

Wind rose: January

(i) State the direction from which longshore drift comes. **[1]**

(ii) What is the prevailing wind direction? **[1]**

(iii) Explain the relationship between your two answers above. **[3]**

(c) With reference to an area you have studied, suggest why and how the movement of beach material can be controlled. **[8]**

[Edexcel (ULEAC) specimen]

(8) The synoptic (weather) map below shows an occlusion over the UK. Study it.

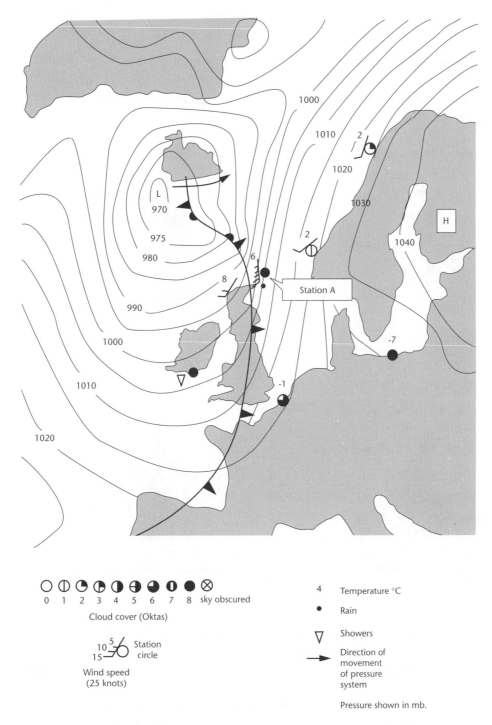

0 1 2 3 4 5 6 7 8 sky obscured
Cloud cover (Oktas)

10 —5— Station
15—⚲ circle

Wind speed
(25 knots)

4   Temperature °C

•   Rain

▽   Showers

→   Direction of movement of pressure system

Pressure shown in mb.

(a) Describe and explain the weather being experienced at Station A.   **[8]**

(b) What changes in precipitation would you expect at A in the next 24 hours?   **[8]**

(c) The passage of a depression over the UK can bring a variety of weather hazards for people. Choose an actual hazard event you have studied and then describe any steps that were taken to reduce its impact.   **[9]**

[OCR, 2000]

(9) Describe and explain the differences between temperate depressions and tropical cyclones (hurricanes).   **[20]**

[Edexcel (ULEAC) specimen]

Mock Exam

Time: 1 hour

Instructions

Answer 2 questions.

(1) Study Figure 1, which shows the actual and projected population pyramids for the most and least developed countries.

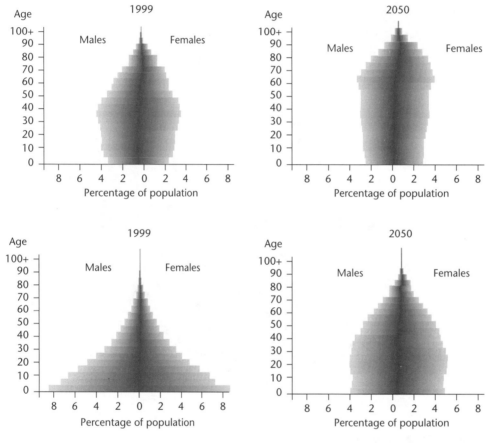

Figure 1

(a) (i) Describe the differences between the population pyramids for year 1999 and the projected figures for 2050 for the most economically developed countries and least economically developed countries. **[6]**

(ii) Choosing either the most economically developed or the least economically developed countries, suggest two reasons for the changes you have outlined in your answer to **(a)(i)** **[4]**

(b) Government policies designed to change population growth rates are either termed pro-natal or anti-natal.

(i) Distinguish between pro-natal and anti-natal policies. **[3]**

(ii) Describe the policies employed by one country in order to reduce population growth. **[5]**

(iii) What have the social and economic impacts of these policies been? **[7]**

(2) Study Figure 2, which shows the process of suburbanisation for an MEDC city.

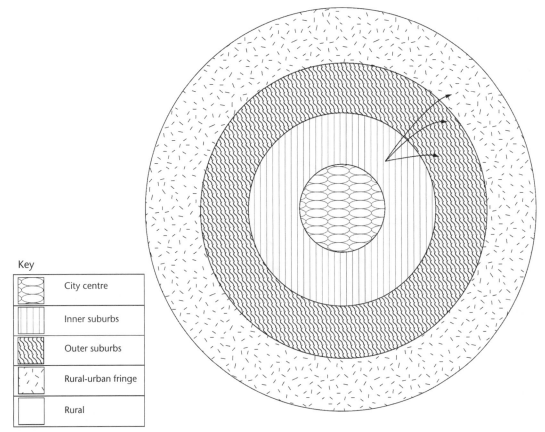

Key

	City centre
	Inner suburbs
	Outer suburbs
	Rural-urban fringe
	Rural

Figure 2

(a) Using located details from one named large urban area in an MEDC:

    (i) Draw on and label any other processes of urban population change onto Figure 2 above. **[6]**

    (ii) Select two of these processes and explain why they are occurring in your chosen city. **[6]**

(b) (i) With reference to the rural-urban fringe, describe and explain the impact of suburbanisation on the physical environment. **[5]**

    (ii) Outline some ways in which you could use primary fieldwork and secondary research to study the changes outlined in **(b)(i)** **[4]**

(c) (i) Define urban regeneration. **[2]**

    For an MEDC city of your choice:

    (ii) Describe one policy that has been designed to encourage urban regeneration. **[4]**

    (iii) Evaluate the impact of regeneration upon the inner city. **[6]**

(3) **(a)** Study Figure 3, which shows changes in the number of employees in manufacturing industry in the United Kingdom between 1978 and 1996.

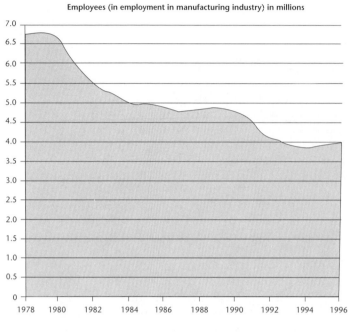

Figure 3

   (i) Describe the changes shown in Figure 3. **[2]**

  (ii) State two reasons for the changes shown under each of the following headings: **[4]**

    1. Domestic (national)

    2. Global (international)

 (iii) Figure 4 gives some additional information about manufacturing industry in the UK.

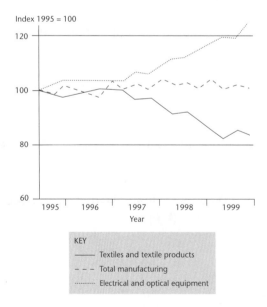

Figure 4

Outline one way in which a different view of manufacturing industry in the UK is shown in Figure 4. **[2]**

(b) Explain the relationships between the percentage working in the secondary sector of manufacturing and the level of economic development of a country. **[7]**

[AQA, 2001]

Answers

(1) (a) (i) The vent through which molten material escapes from the crust.

(ii) Coastal margins, e.g. North West America; mid-oceanic, e.g. mid-Atlantic; others at 'hot spots', e.g. Hawaii.
It is important to mention bands and belts here.

(b) (i) 1. D or E.
2. C.

(ii) Crustal weaknesses above a plume of rising magma, the result of convectional cell activity, allow molten magma to escape through a 'hot spot'.

(c) (i) Formation of batholiths, dykes, sills, etc. You need to focus on sub-surface cooling and the lack of routes through overlying material. Mention also the lack of 'pressure'.

(ii) You should offer a range of benefits here, e.g. building materials, minerals, tourism, fertile soil etc. You need to describe, explain and exemplify to gain the highest marks.

(2) There are three aspects to this question; ensure you address each one. Use case-study material from all over the world (see the AS Study Guide) and ensure that it covers the range of developmental states.
Refer to both primary and secondary effects.
Make reference to the actual impacts of such effects.
Explain how the country manages the effects and the aftermath.

(3) There are two main elements to this question:
1. Detail about the hazard risk in volcanic areas.
2. The processes of population increase, settlement expansion and economic growth in such regions despite the risk.
This question is synoptic; you must mention other pertinent points for whatever examples you use.

Your answer should refer to the risks of:
* Pyroclastic showers.
* Lava flows.
* Earthquake activity and volcanic eruptions.
The importance of tourism, geothermal power and rich volcanic soils as you explain population/socio-economic increases and economic development/advantages is a must. Use examples!

(4) (a) (i) A freshwater succession or hydrosere. Alternatively, you could focus on stages.
Ceases to dominate at stage 5

(ii) The dominant plant is the tree. These grow to full size and then shade out smaller plants. Only those able to stand constant shade survive. Even when there are some trees it is possible for plants to survive. It is important in this question to communicate your ideas on ecosystems effectively.

(iii) It is important to mention size/height here. Some areas are not covered at all by shade; other areas are heavily shaded. You must reference both the land and water plants. Mention also the water-tolerant willows and dry-land species like oak. Also varying number of species.

(b) Physical arresting factors generally include volcanic activity, erosion and deposition by rivers. Human arresting factors include deliberate clearance, burning and settlement, and over-use of the land. Cover both human and physical factors in your answer.

(5) (a) (i) Interception is where raindrops are prevented from landing on the ground by vegetation such as leaves and branches in the tree and shrub layer.

(ii) Throughflow is the movement of water downslope through the sub-soil. It is helped by root systems and by impermeable and permeable rock intersections.

(b) Methods could be either random or systematic. You should demonstrate a clear understanding of whichever method is chosen and link it to the task undertaken.

(c) (i)

- Describe the differences between hydrographs in terms of two or more variables, e.g. lag time, the peaks and the slopes of the limbs.
- You need both to qualify your answer here (using facts and your understanding of hydrology) and to quantify it (using data).

(ii)

- You must show at a basic level the differences between infiltration and interception, perhaps based on simple explanations of the differences between rock types and vegetation.
- You must describe the way in which water moves through the hydrological system.
- The very best answers will describe the role of overland flow and throughflow in relation to P and Q, and interception/infiltration and evapotranspiration in relation to P and R.

(6) River management should aim to avoid excessive riverbank erosion, control flooding, maintain water resources and to negate the effects of pollution.

Advantages:
- Reduces flooding, erosion and over-sedimentation.
- Restores rivers to their original state.
- Restores habitats.
- Enhances the environment.

Disadvantages:
- Increases flooding, erosion and sedimentation.
- Destroys habitats and increases scour.
- Destroys spawning grounds.

(7) (a) (i) 5 km +/− .5

(ii) 15th to 17th January.

(iii) Storm waves or rapid transport. The wind changes direction and therefore the direction of longshore drift changes.

(b)(i) 1. NE.
2. SW.

(ii) The SW wind is not as strong as the NE. Use some figures here and don't forget the idea of prevailing and dominant winds.

(c) Include the need to protect coasts, stop spits growing, aid navigation and help tourism. This answer needs examples, not just the theory! Then explain how it is done – groynes etc.

(8) (a)
- Full cloud cover and rain associated with cloud along an occluded front.
- Strong southerly winds as a result of steep east-west pressure gradient.
- Relatively high temperatures for November linked to southerly air streams.
  In this question you would be expected to display a clear understanding of the link between pressure and front system types and their associated weather.

(b)
- Eastward movement of the system will be rapid because of the very low pressure.
- Rain ceases and the sky clears.
- Wind strength drops, and direction becomes westerly and then northerly.
- Showers are possible in polar maritime air.
- Temperatures drop below freezing in northerly air.
  Just describe the sequence of events for this question. You will get no extra marks for added explanation. Concentrate on changes in precipitation.

(c) Concentrate on one well-described hazard here.
- Strong winds, heavy rain (flooding) and snowstorms are the best routes here. An example from the UK is what is really needed.
- You must attempt to cover the scale and nature of the weather conditions – don't explain them. Things like storm damage, traffic disruption and flooding of farmland will aid your answer.
- Short-term responses and the steps taken to reduce the impact could be mentioned, but what is really required here is a reference to long-term measures, e.g. engineering works or emergency service provision.

(9) You must describe and explain here. Importantly, you must ensure that there are locational comments in your answer.
- Comment on differences in formation.
- Show you know how, why and where they are formed.
- Show you know something of the route that they take.
- Show that there is a degree of seasonality in their formation.
- Diagrams will aid your answer.
- Describe the weather associated with each system.
- Briefly comment on effect on humans.

Answers

(1) (a) (i) Use descriptive terms such as base, top, wide, narrow, concave, convex, straight and tapering to describe the changes, but be sure to include some data in your answer.

(ii) There are a number of possible answers here, but you need to fully outline the reasons behind your suggestions:
Most developed countries – later/no marriages, career women/men, desire to spend money on consumer goods and recreation, improving healthcare – longer life expectancy.
Least developed countries – improved healthcare including advancements in post- and pre-natal care, better access to contraception, better education about contraception, government policies to reduce birth rates, societal changes and the rise of consumerism, more opportunities for women, improved nutrition.

(b) (i) Use examples here and stress that **anti-natal** policies are designed to reduce births, e.g. China, and **pro-natal** policies are designed to encourage births, e.g. Singapore. For three marks you should also suggest why this might be the case.

(ii) Be specific and thorough about your case study. If you use China as an example, you will need to have excellent knowledge as many students use this case study. A better example might be lesser known but more individual.

(iii) Include a balance between the **positive** and **negative** impacts, as well as **social** and **economic** ones. It often depends on how the policy has been enforced as to the social implications, e.g. in China some women have been reportedly forced to have abortions of second children and how efficiently it has been enforced as to the economic implications, e.g. the Chinese economy is booming.

(2) (a) (i) Be sure to make specific reference to a large urban area, somewhere with at least 100,000 people. Label the map with examples of places within your urban area that are exhibiting population change.
- Suburbanisation (outward movement within an urban area).
- Urbanisation (rural to urban migration).
- Counterurbanisation (urban to rural migration).
- Reurbanisation (movement back into urban areas).

(ii) Reasons:
- As lower income families earn more money.
- As people look for employment out of agriculture into better paid jobs.
- Wealthy return to countryside – car ownership enables this.
- Upper-middle income groups move to renovate character houses near city centre amenities.

(b) (i) Suburbanisation usually causes urban sprawl and thus you should consider the impact of things such as housing, utility provision, communications, traffic and waste disposal on hydrology, soils, open land, farmland and air pollution.

(ii) You are only required to outline here, but should include a balance of ideas between primary and secondary data collection techniques. Primary data collection might include: Environmental Impact Assessment, Environmental survey, Questionnaires of local residents. Secondary data collection might include: OS mapwork, Local newspaper article search, Local planning office plans.

(c) (i) Urban regeneration is literally the act of giving new life or vigour to an urban area, which will include the redevelopment of older and run-down urban environments.

(ii) Your policy could be either to discourage development elsewhere in the urban area, e.g. greenbelt policy, or to encourage development in the inner city, e.g. local government incentives offered to redevelop certain sites.

(iii) Evaluate is a high-level term and thus you need to reach a conclusion about regeneration, balance your argument and include positive impacts, e.g. gentrification of older urban areas, and negative impacts, e.g. high cost of the redevelopment projects.

(3) (a) (i) Comment upon the decline using figures, e.g. 1978 – 6.5 million employees; 1996 – 4 million employees, and also upon the different rates of decline, e.g. 1980 to 1984 – rapid decline from 6.6 to 5 million.

(ii) There are a whole range of reasons to quote here. You could include:
- domestic: mechanisation, rationalisation, deindustrialisation;
- global: competition from cheaper imports, movement of manufacturing plants to cheaper labour cost locations, increasing role of the TNCs.

(iii) The graph is an index, and therefore shows the amount produced in 1995 as 100. Any figures after that are proportional to the 1995 figures. You could include ideas such as:
- manufacturing output has remained static, even though the employment has declined;
- some industries have grown, but it is industry-specific.
  Remember that the graph in Figure 1 starts in 1978, so you are not comparing the same time spans.

(b) Here you should understand the transition of a country from a traditional to a post-industrial economy, i.e.
- traditional economy – high primary and low secondary employment levels;
- industrial economy – secondary employment increases at the expense of the primary;
- post-industrial economy – tertiary employment increases at the expense of the secondary.
  Remember that this all ties into the needs of the country, and you should therefore include examples of countries in your answer.

Mock Exam

Time: 2 hours

Instructions
Answer 2 questions in Part A and 2 questions in Part B.

Part A

(1) **(a)** Explain the difference between:

    (i) a valley glacier and a continental ice sheet **[3]**

    (ii) snow accumulation and ice ablation **[3]**

    (iii) the nature of glacial and of fluvioglacial (meltwater) deposited materials **[3]**

  (b) Choose either valley glaciers or continental ice sheets.

    For the one chosen, outline the ways that its work has both restricted and provided opportunities for human settlement. **[6]**

[AQA (AEB) specimen]

(2) The diagram below attempts to model some of the aspects of desertification. Study it.

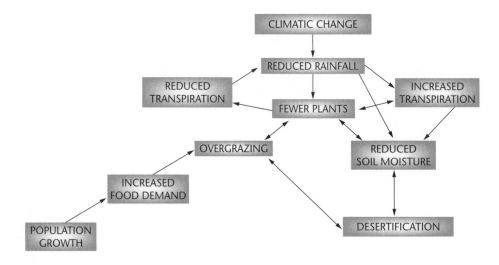

  (a) **(i)** State two immediate causes of desertification shown on the model. **[2]**
    (ii) Explain how each cause originates. **[6]**
  (b) Attempt a definition of desertification. **[4]**
  (c) In what present-day climatic type does desertification currently take place? **[1]**
  (d) Explain the interaction between increased evaporation, reduced soil moisture and fewer plants. **[6]**
  (e) Suggest three changes in agricultural practice or technique that might reduce the rate of the desertification process. **[6]**

[Edexcel (ULEAC)]

(3) Study the diagram below.

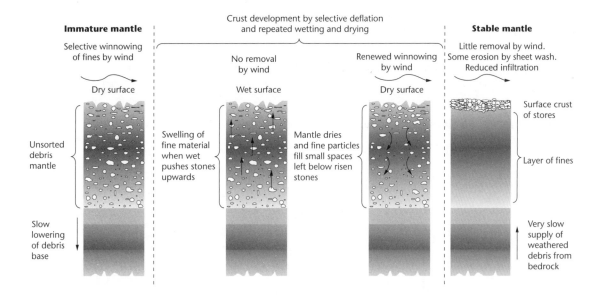

Source: Bishop, V, and Prosser, R, *Landform Systems*, Collins Educational, 1997

(a) Explain how desert pavements are formed. **[4]**

(b) Why are desert pavements so vulnerable to damage by vehicles and other factors that disturb the pavement? **[4]**

(c) Discuss the role of water in the formation of desert features. **[6]**

Source: Franks, P, and Guiness, P, *People and the Physical Environment*, Hodder and Stoughton, 1998

Part B

(1) The impact of natural disasters owes more to human than to physical factors. With reference to specific examples, discuss this statement. **[25]**

(2) (a) Compare the changes in organic and inorganic matter with depth as shown in the diagram below. **[5]**

(b) Examine the influence of physical and human factors on the profile of the soil. **[20]**

[Edexcel]

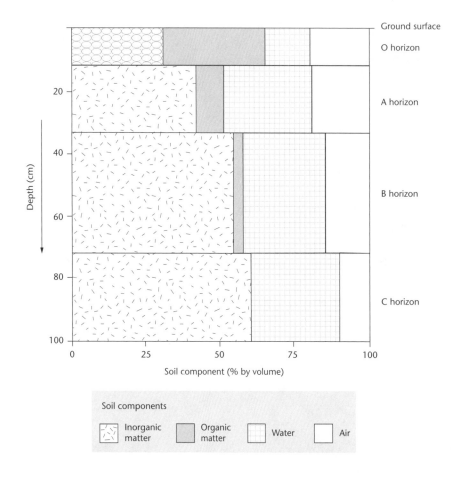

(3) Identify the principal causes of changes in the relative levels of land and sea, and examine the effects of such changes on coastal morphology. **[20]**

(4) Discuss the social and economic impacts of the late arrival of the summer rains and also the risk of flooding and storm surges, in a country of your choice. **[20]**

(5) Explain and exemplify how the same glacier can both erode and deposit. **[20]**

(6) 'Forest management schemes are a mixed blessing.' Discuss with reference to areas you have studied. **[25]**

(7) Describe and explain the reasons for the aridity of named desert environments. **[25]**

Mock Exam

Time: 2 hours

Instructions
Answer *2* questions in Part A and *2* questions in Part B.

Part A

(1) 'International migrations have changed a great deal in the last 30 years but these changes have had little effect upon the issues that arise in multi-cultural societies which have generally stayed the same.'

Discuss the extent to which you agree or disagree with the above statement.
In your answer you should:
- identify the major forms of international migration
- indicate how the major forms of international migration have changed or are changing
- outline the issues which arise in multi-cultural societies
- assess how the changes in migration are affecting the issues that you have identified. **[25]**

[AQA Specimen]

(2) Study the data shown in Table 1 for a variety of countries at different levels of economic development.

Country	GNP per capita (US$) 2001	Adult literacy rate (%) 2001
1. Japan	40,940	99
2. Belgium	26,440	99
3. Argentina	8,330	96
4. Oman	4,950	75
5. Mexico	3,670	90
6. Botswana	3,020	70
7. Tunisia	2,030	67
8. Philippines	1,160	95
9. China	750	82
10. Mali	240	95

Table 1

(a) (i) On graph paper, plot the GNP per capita against the adult literacy rate using data from Table 1 above. Use the number key for each country, and draw in a labelled best-fit line by eye, with a ruler. **[7]**

(ii) Outline the advantages and disadvantages of using a scattergraph and best-fit line as opposed to a Spearman rank correlation coefficient calculated using the same data. **[5]**

(iii) Explain the nature of the relationship as revealed in the graph you plotted in **(a)(i)**. **[5]**

(b) Comment on the view that economic indicators are the only true measure of quality of life in different parts of the world. **[8]**

Figure 1: National Parks in Texas, USA

Figure 2: Padre Island National Seashore, Texas, USA

This National Park encompasses 133,000 acres of America's vanishing barrier islands. It is the longest remaining undeveloped barrier island in the world. White sand beaches, interior grasslands, ephemeral ponds and the Laguna Madre provide habitat for coyotes, waterfowl, reptiles and amphibians, nesting sea turtles, ground squirrels and snakes. While providing food, water and shelter for a multitude of diverse wildlife, the island remains a Mecca for tourists. From sunbathing to windsurfing to fishing, the island provides recreational opportunities for everyone needing to feel the wind in their faces and the surf on their feet.

Figure 3: Padre Island National Seashore information

(3) (a) (i) What improvements could be made to Figure 1 to provide better information for tourists? **[3]**

   (ii) Describe the situation of Padre Island National Seashore. **[3]**

   (iii) Using Figure 3, outline the primary tourist resources that Padre Island National Seashore has to offer. **[4]**

Auto touring	Backpacking	Biking	Bird watching
Boating	Camping	Fishing	Hiking
Horseback riding	Kayaking	Nature walks	Scuba diving
Snorkelling	Stargazing	Swimming	Wildlife viewing
Windsurfing			

Figure 4: Recommended activities in Padre Island National Seashore

(b) Assess the sustainability of three of the recommended activities in Padre Island National Seashore, Figure 4. **[6]**

(c) Evaluate the costs and benefits of increasing tourist awareness of National Parks such as Padre Island National Seashore. **[6]**

(d) Suggest strategies that tourist managers could put in place to manage large numbers of tourists in selected key localities in National Parks. **[8]**

(5) Study Figure 5, which shows John Betjeman's adaptation of the Harvest Hymn.

We spray the field and scatter
The poison on the ground
So that no wicked wild flowers
Upon our farm be found.
We like whatever helps us
To line our purse with pence;
The twenty-four-hour broiler-house
And neat electric fence.

All concrete sheds around us
And Jaguars in the yard,
The telly lounge and deep-freeze
Are ours from working hard.

We fire the fields for harvest,
The hedges swell the flame,
The oak tree and the cottages
From which our fathers came.
We give no compensation,
The earth is ours today,
And if we lose on arable,
Then bungalows will pay.

Figure 5: Harvest Hymn from Collected Poems by John Betjeman,
John Murray (Publishers) Ltd., 1980

(a) Describe the environmental impact of modern farming practices employed in MEDCs as highlighted by the Harvest Hymn. **[10]**

(b) Evaluate the statement 'only farmers in more economically developed countries have the luxury of being able to consider a change to organic farming practices'. **[15]**

(5) Study the table, which shows population data and projections to compare the differences between more and less economically developed countries.

	Population (millions) 2000	Natural increase (annual %)	'Doubling time' in years at current rate	Projected population (millions) 2025	Projected population (millions) 2050
World	6,067	1.4	51	7,810	9,039
MEDCs	1,184	0.1	809	1,236	1,232
LEDCs	4,883	1.7	42	6,575	7,808

Assess the implications of achieving sustainable development in the light of the evidence of the data from the table. **[30]**

Part B

(1) (a) It is claimed that the principal cause of the global development gap is excessive population growth in the less economically developed world. Outline the reasons for this claim. **[15]**

(b) To what extent have strategies been successful in reducing this development gap ? **[10]**

[AQA (AEB) specimen, 2000]

(2) With reference to specific large urban areas, discuss the idea that urban growth is always at the expense of the physical environment. **[20]**

(3) Study Figure 6, which shows the top 10 recipients of direct foreign investment 1985–95 in billion US$.

Country	Direct foreign investment by TNCs 1985–1995 billion US$
USA	477
United Kingdom	190
France	138
China	130
Spain	90
Benelux	72
Netherlands	68
Australia	62
Canada	60
Mexico	44

(a) With reference to such inward investment, what are the costs and benefits for the recipient countries? **[10]**

(b) Assess the range of development options open to the decision-makers in LEDCs who are not receiving such large amounts of direct foreign investment. **[15]**

Figure 6

(4) Study Figure 7, which shows origin of asylum applications lodged in Europe and North America between January and September 2000.

(a) Using examples from Figure 7, discuss the principle reasons why refugees flee from their country of origin. **[12]**

(b) Outline the problems that can befall a refugee on the route to gaining asylum in a safe country. **[8]**

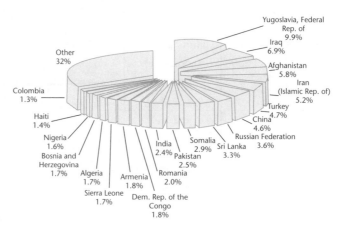

Figure 7

Source: UNHCR Report on Asylum Applications

(5) (a) Outline the potential reasons why
food aid is provided. **[7]**

(b) Discuss the contention that food aid
to less economically developed countries
is only a short-term solution
to a long-term problem. **[18]**

Figure 8: Mariamma Ali and Abdi collecting their food ration
from an Oxfam supplementary feeding centre,
distributing food aid in the Horn of Africa

Source: Crispin Hughes/Oxfam

(6) Study Figure 9, which shows Doxey's
index of host irritation.

With the aid of specific examples from LEDCs,
comment on the usefulness of models of tourism
development, such as Doxey's index of host
irritation, in planning for future tourist
development. **[25]**

Figure 9   *Source*: John Edmonds,
The Geography Collection: At Leisure,
Hodder and Stoughton

(7) Examine the view that with increasingly efficient agricultural techniques being
employed in more economically developed countries, the environment will not be
the only thing that will suffer. **[25]**

(8) Sustainable tourism projects such as ecotourism are seen to have positive
environmental impacts. Using examples from a range of different scales, evaluate
the socio-economic impacts of sustainable tourist projects. **[25]**

Answers

Part A

(1) (a) (i) A valley glacier is a glacier which occupies a pre-glacial valley, either formed by the coalescence of several cirque glaciers or formed at the edge of an ice sheet. An ice sheet is a large, continuous layer of land ice. Considerably thicker, it may cover regions on a continental scale. You could also mention location, size and extent.

(ii) Accumulation = build-up of snow and ice by precipitation; ablation = ice and snow melting. You could develop this answer along the neve route!

(iii) Boulder clay versus sand and gravels. Develop this answer along the sorted and unsorted/rounded and angular route.

(b) There are two aspects to this question – restrictions and opportunities.
Include detail on:
Valley glaciers – restrictions: sharpen the landscape, form steep slopes, impede transport; opportunities: big valley floors for settlement, lakes for water supply, hydro-electric power and tourist sites.
Ice sheets – restrictions: remove all topsoil and create lots of lakes; opportunities: boulder clay on the ice margin is good for farming.

(2) (a) (i) Overgrazing and reduced soil moisture are possible answers. There are other causes of desertification you could choose.

(ii) Increased food demand/rising populations; climate change.

(b) Desertification is the extension of typical desert landscapes where they did not occur in the recent past and is 'produced by a combination of increasing human and animal populations, causing the effects of drought years to become progressively more severe so vegetation is placed under stress'.

(c) Sub-tropical climatic zone.

(d) This answer should be an explanation of the contribution climatic change makes to desertification.

(e) • Reduced deforestation.
• Restricted grazing.
• Reduced cultivation.
• Return to traditional cultivation methods.
• Change of energy sources.
• Corralled animals.

(3) (a) Pavements develop where:
• Vegetation is absent.
• Water is shed quickly.
• There is a supply of coarse material/debris.
• Veneers develop; these tend to be impermeable.
• There is time for pavements to develop.
• Fine material is removed by wind causing remaining pebbles to settle and interlock like cobble stones.

(b) Once disrupted, pavements don't heal quickly. Accelerated erosion occurs when the surface patina is lost, usually to off-road vehicles.
Desert pavements are vulnerable because:
- There is no soil cover.
- Patinas are thin.
- Wind and water play a major part once the patina is lost, exploiting cracks and joints.
- Increasing numbers of people visit these areas.
- There is increased use of marginal lands near deserts by nomadic farmers.

(c) Processes: water occurs rarely in desert areas. Rainfall is intense, short-lived and localised, releasing high energy for a short time and causing intense erosion and etching. As the rain decreases, there is a lot of sediment to distribute.
Features include: arroyos/wadis, bajadas and playas. Water can also cut pediments.

Part B

(1) You need to show an understanding of the connections between, for instance, plate tectonics, hydrology, slopes and settlement. Your answer must show synoptic understanding. You must consider how physical factors cause natural disasters and how human factors, like population density, hazard perception and development link in. Use examples and ensure you attempt some synthesis of the significance of human and physical factors.

(2) (a) The principle of this question is to recognise that the organic material is decreasing and the inorganic material is increasing. You should recognise in particular the absence of organic material in the C horizon. You must compare the horizons and profiles.

(b) • You should cover a range of human and physical soil-forming factors.
- Make sure that at least two soils are referenced in your answer.
- Make sure there is some appreciation of time with regard to soil formation in your answer.
- You should write about the appearance and character of the profile.
- You could usefully draw a soil profile and label it.

(3) The two causes of sea-level change are:
- Eustatic changes, i.e. worldwide changes in absolute sea level possibly due to climatic change.
- Isostatic/tectonic changes, i.e. changes due to the redistribution of weight on the land surface.

Morphological effects are:
- Submergence, e.g. into rias and fjords.
- Emergence, e.g. raised sea beaches and barrier islands form.
Also mention factors such as increased erosional range, coastal flooding, etc.

(4) This question requires an answer that looks at both social and economic impacts, and the extremes of rainfall, i.e. too much and too little. It also requires an understanding of why rains arrive late. Being able to explain why there are floods and storm surges is implicit in this question.

Overall, there is a range of ideas here, but impacts include:
- 80% of annual rainfall arrives with the monsoon; it affects a quarter of the world's population.

Agriculture and impacts

- Distribution and seasonal incidence of rainfall is important.
- Drought in the 1960s reduced rainfall; drought affected crop yields, which fell by 80%. This all led to starvation. In 1987 the summer crop and five million cattle were lost, 3.5 million tonnes of food aid was donated and a further £150 million of aid was given.
- In the 1970s rainfall was up and yields increased.
- Two crops can be grown when rainfall is high. High yields mean that everyone is well fed.

Hazards and impacts

- Flooding is the major hazard (1993 was a heavy monsoon year). July 1987 brought 12 months of rainfall to South-East Asia; 12 million people were affected, 2 million tonnes of food lost and 20,000 km of roads damaged. Eventual costs to India alone were nearly £1 billion. Very little aid was given by MEDCs.
- Deaths are high during monsoon times, road and rail links are cut, farmland inundated, power stations disabled and landslides initiated.
- The scale of the flooding hazard is such that countries like India have to adjust their way of life instead of trying to prevent the effects of the monsoon.

(5) You should show an understanding of how most erosion occurs in the upper section and how most deposition occurs in the lower part of the valley glacier. The two processes depend upon the slope/gradients, rates of accumulation and ablation, i.e. the glacial budget.

Glaciers are dynamic, eroding material in one place and transporting and depositing it in other places.

Use present-day examples: the glaciers of the high Alps and of Canada are ideal.

(6) You could choose either a local or a more distant area to study. In other words, your local wood is as relevant as the Brazilian rainforest.

Focus on:

- The competition that exists between the users of such forests.
- The gains and huge losses to be made from the use and abuse of forest areas, wherever they might be.
- Variations in the use and abuse of forest areas over time.
- The environmental, economic and social nature of the gains and losses.

(7) Causes of climatic aridity include:

1. The nature of the relief, and the distribution of land and sea.
2. The nature of atmospheric circulations, trade winds and associated sea currents.

The Sahara, Namib, Thar and Kalahari deserts, and California's and Australia's deserts, are all caused by climatic aridity.

Other desert environments are the result of:

1. Desertification, initiated by man, e.g. the Sahel.
2. Shifts and changes in temperature and temperature belts, and latitudinal shifts in continents in the longer term.

Answers

Part A

(1) Focus your essay on:

- **The major forms and changing levels of international migration** – i.e. voluntary or forced, temporary or permanent, within and between LEDCs or MEDCs. For example:
  1. voluntary and permanent, MEDC to MEDC – decreasing except for examples such as within the EU with the free movement of labour;
  2. forced – refugees as a result of famine have increased due to population pressure, increasing conflict in Sub-Saharan Africa and the changing climate.
- **The issues** – cover at least two issues in depth, e.g. housing, employment, religion, food, dress, education, racism.
- **How the changes are affecting the issues**, e.g. large influxes of refugees put strains upon the host countries' resources; if migrants accept low-paid jobs they may be keeping the pay levels low for the indigenous population; the pressure of political refugees for services may cause resentment within the host community.

All of the ideas should be linked and the structure should be clear as it has been given to you at the outset of the question.

(2) **(a) (i)** Plot the graph accurately and neatly, and remember to include the following points:

- GNP pc should be along the **x-axis** as it is the **independent variable**.
- Include a number next to each point to identify it.
- Label the axes clearly and accurately.
- Roughly half of the points should be on either side of the best-fit line.

(ii) The advantage of the graphical technique is the clarity of the relationship as long as there are not too many anomalies. The statistical test gives you statistical certainty but if the calculated coefficient is not significant, no trends can be seen.

(iii) The variables have a **positive relationship**; include **data** and examples to support this and any **anomalies**. This positive relationship is due to a higher GNP pc, enabling higher investment in education and therefore better literacy rates. There are exceptions, and you could suggest reasons for this, which might include ideas such as targeted development programmes and variations in data collection techniques.

(b) This is a common development question, which requires you to look at the range of development indices:

- **economic**, e.g. GNP pc;
- **demographic**, e.g. crude birth rates, life expectancy;
- **social**, e.g. literacy rates, calorific intake;
- **other measures**, e.g. PQLI, PPP, HDI.
  Economic indices tend to ignore regional variations, and do not take into account the cost of living in the country or the influence of the informal economy. A balance is needed, and the measures that take into account a number of factors are often best.

(3) (a) (i) Features such as a clearer key, a linear or ratio scale, physical features such as high land or mountains, and labelled rivers, seas, oceans and neighbouring states.

   (ii) Situation is what is asked for, that is where it is in relation to other places, e.g. south-east of the state, coastal, at the southern limit of Highway 37 from San Antonio, just north of the Palo Alto Battlefield National Park.

   (iii) Primary tourist resources are the 'natural' ones. Reading Figure 3, they are: the aesthetic beauty of the barrier island itself; the white sandy beaches; interior grasslands; habitats for a variety of mammals, birds and reptiles; the climate suitable for sunbathing; and the ocean for water sports and fishing.

(b) Remembering the definition of sustainability in tourism, choose those activities that are clearly sustainable, e.g. stargazing, and those that are non-sustainable, e.g. auto touring. Consider the impact of these activities based upon your answer to the previous question.

(c) The ultimate National Park aims to promote public access, awareness and understanding (the basis of sustainability), but also to limit numbers so as to preserve the park for future generations.

(d) Choose a simple idea such a honeypots, and use examples that you have studied to show how they operate.

(4) (a) • First, identify the **modern farming practices** from the text:

   fertilizer and pesticide application; battery farming of livestock; the removal of traditional fences and hedges and the use of electric fences; stubble burning; the replacement of older farm buildings with functional ones using non-traditional materials.

   • Then consider their potential **environmental impacts**:
   pollution of groundwater, eutrophication, air pollution, removal of natural habitats, reduction of biodiversity, visual pollution.

   • Try to balance this with some consideration for the modern agribusiness.

(b) This question needs to be addressed with specific case-study material, and needs to consider farmers from both MEDCs and LEDCs. Consider the structure of the farming industries, and the commercial and sustainable aspects of production as well as the farmers' ties to the land. You should consider the following ideas:
MEDCs – financial support for conversion available, low returns during two-year conversion, commercial implications, higher returns from niche market upon conversion, maintenance of natural environment.
LEDCs – inability to afford artificial fertilizers may necessitate the use of organic fertilizers, lack of market for specifically organic crops, possible perception that fertilizers and pesticides are a necessity to maintain production.
You could consider the sustainable angle and thus whether converting to organic practices is ultimately a luxury or a necessity.

(5)   The table reveals the huge demographic differences between more and less economically developed countries, such as the already high population of the LEDCs, the differences in rates of natural increase and the consequent disparities in doubling time for the population and projected populations.

The emphasis of the answer is therefore on the relationship between population and resources and the pressure that such huge population increase will place upon LEDC resources. Include some theory here, e.g. Malthus, Boserup and the ideas of over- under- and optimum population.

Sustainability can only be achieved by:

- managing the population growth;
- using the population as a resource to develop the economy;
- developing sustainable resources;
- promoting sustainable economic growth.

Include the different approaches to sustainability, i.e. top-down and bottom-up schemes, and the societal perceptions of sustainable development, e.g. from fashionable idea to environmental necessity.

Part B

(1) (a)   Population growth is the focus, there are pressures (policies) which are reducing the rate; problems include money for development diverted for services, lack of job opportunities leading to migration, pressure on farming; and other factors include activities of Trans National Companies, political turmoil, shortage of capital, government policies, neo-colonialism.

(b)   Argue the case, reaching a conclusion based upon ideas such as: rapid industrialisation; tourist development; environmental problems and other costs; agricultural development mechanisation – job losses; aid – good but tied to dependency; family planning policies. Use examples to aid your case.

(2)   You should aim to use at least two contrasting examples of large urban areas, e.g. Cairo, Sao Paulo, and consider how urban growth or sprawl might affect: ecosystems – loss of farmland, loss of biodiversity; hydrology – river pollution, changes to local hydrology; atmosphere – atmospheric pollution.

Show that you understand the physical systems that you are writing about, e.g. the impact that urban environments have on interception rates, surface run-off and eventually discharge. Also discuss the issue, giving a balanced viewpoint that will enable you to write an interesting conclusion. This might include some suggestions about future planning or sustainability.

(3) (a) There are a number of costs and benefits to the economies of **MEDCs**, so make sure to cover aspects of:
capital; degree of regulation; dependency; employment and labour; industrial structure; technology; trade and linkages.

(b) The focus here is not on 'top-down' development, which might see development as merely increasing the GNP/capita, but on 'bottom-up' development, which serves to improve the **quality of life** of the population. Try to use a range of examples, and include ideas from industry, tourism, agriculture, education, health, housing, amenities and sanitation.

(4) (a) A definition of terms could be the route into this essay question; a refugee is a person who is unwilling to return to her or his homeland for fear of persecution based on reasons of race, religion, ethnicity, membership of a particular social group or political opinion. You could consider each aspect separately, and include some of the following examples:
religion, e.g. the persecution of Muslims in Northern India; **political**, e.g. people dispossessed due to the conflict between the state and left-wing guerrillas in Colombia; **ethnicity**, e.g. Kurdish persecution in Turkey and Iran; **political opinion**, e.g. anti-government supporters in Nigeria.
Use these terms to generate examples from the pie chart, but make sure to recognise that migration streams are made up of many different individuals who may have complex reasons for fleeing.

(b) Make sure to include a number of the following ideas and include examples from your studies and reading: restricted movement, limited border access, forced repatriation, non-recognition of refugee status, lack of legal protection, alienation, xenophobia, racism and physical barriers.

(5) (a) Food aid can take the form of free food being provided, or food subsidies being applied to make staple foods more affordable. There are a number of reasons for giving food aid, and you should be aware of this from your studies, but they can be divided in the following way:

- **transparent** – humanitarian relief in a crisis or until the country can feed itself again, e.g. food aid given to Honduras in the wake of the destruction caused by Hurricane Mitch in 1998; morally rich countries should help poorer countries, e.g. the largest food donors are USA, EU, Canada, Japan and Australia
- **hidden** – provided for self interest: to provide aid to create a market for surplus produce, e.g. wheat given as food aid in West Africa during the 1980s, or to support a friendly government or to withhold it from an unfriendly one, e.g. sanctions imposed upon the Cuban state to withhold aid and stop external trade.

Remember also that there are different methods of delivery: charities, multilateral and bilateral and each of these methods of delivery has its own agenda.

(b) It is essential here that you establish the difference between **short-term or emergency aid** and **long-term or development aid**. Food aid is part of the short-term solution, and you should discuss the reasons behind it not being good in the long term, e.g. dependency – postpones economic dependency and self-sufficiency, undermining of local production – depressing local food prices, ultimately the poorest people are the last to receive aid and the fact that it can be used as a bargaining chip in conflicts.

(6)   This is a long question but one which is specific about **LEDCs, models and planning for future development**. You could use a number of different models:

- **Doxey's index of host irritation** – enable planning to take place to promote better host/visitor relations which would incorporate land-use planning, transport planning, marketing and regulations to promote sustainable development, e.g. the Dominican Republic where huge increases in visitor numbers have highlighted the gulf between the wealthy tourists and relatively poor island inhabitants, which has led to a rise in crime levels and warnings for tourists about straying off the beaten track.
- **Butler's model of evolution of tourist areas** – to understand the numbers and types of tourist that could come, and anticipate or head off development, stagnation, decline and rejuvenation, e.g. Thailand recognises stagnation in some areas and is thus rejuvenating itself with the promotion of more sustainable and cultural tourism attractions.
- **core periphery enclave model (Chrispin and Jegede)** – to understand the international social and economic influences of tourism upon the host society and economy, e.g. Costa Rica has developed a sustainable tourist industry that is principally managed by local people and hence the money is reinvested to improve the resources and preserve the cultural heritage of the people.
- **Plog's psychographic positions of selected destinations** – helps us to understand what type of tourist might come from what particular country, depending on accessibility, familiarity and security, e.g. due to the security problems in Colombia, it caters for hardy independent travellers from MEDCs and businesspeople from all around the world. To open up its tourism market, it must improve security and access.

You might even use Myrdal's process of cumulative causation to look at the multiplier effects that could be experienced with growth in tourism.

(7)   You could begin by defining what the efficient agricultural techniques are, e.g. larger fields and the destruction of hedgerows, increased use of pesticides and fertilizers, and what impact they have on the environment. Consider both the direct and indirect consequences of the techniques, e.g. loss of hedgerows means a reduction in biodiversity, but the use of pesticides may have an impact upon a water course, as well as the long- and short-term impacts. You are then open to evaluate the impact such techniques have on humans from the perception that the 'countryside' is losing its attractiveness, to the build-up of chemical residues in our bodies as a result of the use of chemicals in the production of our foodstuffs. You should then make sure you bring into your essay any suggestions that this is a situation that might be changing, like the increases in organic production, or consider a wider context by comparing it to the less economically developed world.

(8) Make sure to consider other examples of sustainable tourism, such as green tourism, responsible tourism, soft tourism, agro tourism, and cover a range of different management styles. There are a whole range of examples to use which have different social and economic impacts (make sure you consider positive and negative):

- **Scuba diving in Belize** may only have minimal human contact and therefore little social impact, but may mean that the economic returns are concentrated on fewer individuals.
- **Trekking in Nepal** involves much greater human interaction and has led to a loss of cultural identity, but the economic returns have been more widely distributed.
- **Rafting on the Colombian Amazon** is well managed but the need for access has meant the construction of airstrips into the rainforest and therefore a loss of isolation and identity due to increased communication and trade with the capital, Bogotá.